More Mysteries of Science

Research Activities for Investigating Scientific Fact and Fiction

by
Tom Christie, Ed.D.

illustrated by Gary Hoover

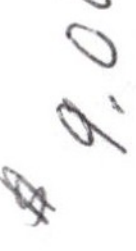

Cover by Gary Hoover

Copyright © 1994, Good Apple

ISBN No. 0-86653-820-8

Printing No. 987654321

945584

Good Apple
1204 Buchanan St., Box 299
Carthage, IL 62321-0299

Paramount Publishing

Table of Contents

How to Use This Book

While *More Mysteries of Science* was designed and written with the academically talented in mind, it can be effectively used to stimulate advanced level thinking skills in all students. The development of activities and extenders is based on the use of higher level thinking skills (Bloom) and processing skills.

Research indicates that by the year 2000, the knowledge base will double every six months. How is it possible to believe that as educators we are capable of teaching that type of knowledge base? We cannot. We can, however, teach our students how to learn. We can begin to renew excitement in older students by stimulating their divergent thinking skills and generating in them a "knowledge" of processes which will afford them opportunities to discover and learn about the world in which they live–opportunities that will help them think for themselves and apply that knowledge into all aspects of their own lives.

Each chapter contains several activities which the student may wish to complete following the reading of the subject matter. These activities provide a wide range of skill development and enhancement and are cross-curricular. Knowing that immediate feedback is important to all students, not just the academically talented, most of the student activities are self-checking. The students know right away whether or not they have been successful. If not, they can easily go back and correct any mistakes.

Each unit concludes with a list of learning extenders. These activities have been written using the various levels of Blooms Taxonomy of Educational Objectives. The activities also incorporate processing skills when appropriate. (Processing skills include observing, classifying, inferring, measuring, predicting, and communicating.) Through these efforts it is hoped that students will be stimulated to develop appropriate problem-solving techniques and critical-thinking skills; learn to solve problems in divergent ways rather than looking for one right answer; and expand their "classroom" to include their home, community, and world.

However, do not limit yourself simply to the extenders and activities presented for each subject area. Using Developing Extenders for Academically Advanced Students at the back of this book (page 55), work with your students to personalize their study of these scientific mysteries. You are encouraged to take sufficient time to discuss the purpose of each activity prior to the students beginning their study. Encourage students to share their own viewpoints, thus motivating them to apply their learning in a variety of other life encounters.

Every effort has been made, at the time of publication, to insure the accuracy of the information included in this book. We cannot guarantee, however, that the agencies and organizations we have mentioned will continue to operate or to maintain these current locations indefinitely.

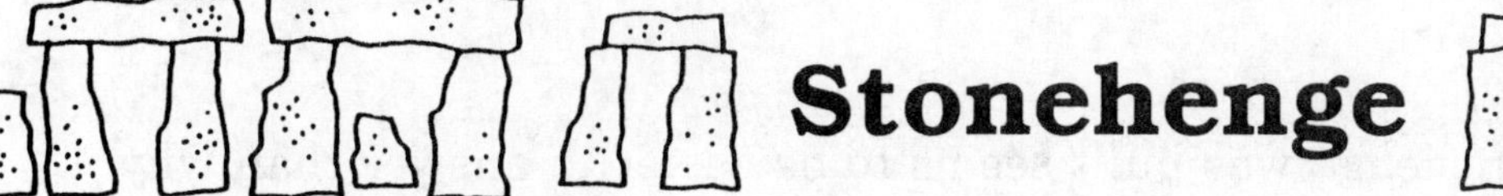 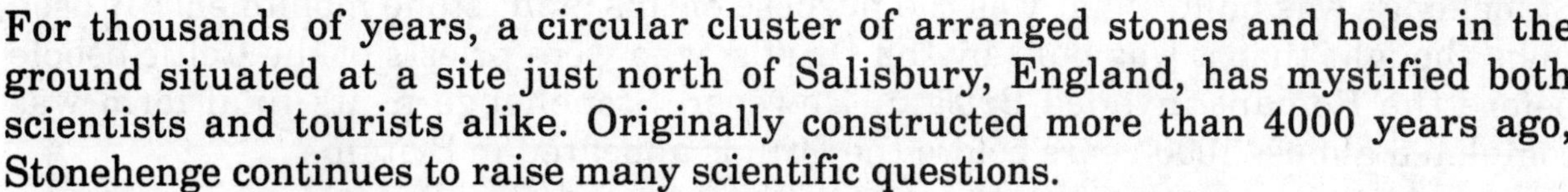

Stonehenge

For thousands of years, a circular cluster of arranged stones and holes in the ground situated at a site just north of Salisbury, England, has mystified both scientists and tourists alike. Originally constructed more than 4000 years ago, Stonehenge continues to raise many scientific questions.

It seems impossible that people living in 2500 B.C. would be able to shape and move these giant stones without modern tools such as drills and cranes. But as impossible as it seems, they were able to do it. Who built it? How did they do it? Why did they do it? We have some of these answers thanks to the work of Professor Gerald Hawkins of the Smithsonian Astrophysical Observatory and his computer.

Scientists believe that the construction of Stonehenge was completed at three different times. The first Stonehenge was made 200 years before the Egyptians started work on the pyramids. A farming people who were skilled in the use of timber made religious buildings in the form of circles with great wooden posts. The sacred area was enclosed by a bank surrounding a ditch. Inside the bank, there was a series of fifty-six small holes or pits.

The second phase began around 2100 B.C. by a people known as Beaker Folk. (They were called Beaker Folk because beaker-like vessels have been found in their graves.) Around the center of the Stonehenge enclosure, they built a double circle of eighty upright stones known as bluestones. Scientists have concluded that the stones were brought all the way from the west of Wales–130 miles from Salisbury. It is theorized that these stones–many weighing several tons–were carried most of the way by water since they had no wheels or pack animals. The Beaker Folk also built a wide roadway now called the Avenue. This 65-foot road was marked by a low bank and ditch on each side and led from the main sacred enclosure all the way to the River Avon which was two miles away.

The third building phase lasted from about 2000 B.C. to 1100 B.C. and was carried out by early Bronze Age people. They removed the bluestone circle and erected a ring of about thirty blocks of sandstone (also called sarsens), some weighing fifty-five tons. These blocks came from a town twenty miles away. How did they get these heavy rocks to the construction site? Professor Hawkins hypothesized that they must have been moved on oak logs used as rollers. He calculated that 800 men would have been needed to haul one of the giant stones. Another two hundred would have had to clear the route and move the rollers. Overall, he has estimated that the entire monument would have taken a total of about 1,500,000,000 working days to construct and involved about 1000 workers at a time!

Once the sandstones were in place, they were linked together by stone lintels. Finally, inside this great sarsen circle, five sets of linteled stones were set up in the form of a horseshoe. The open end faced northeast toward the Avenue, and the closed end contained an enormous sarsen now known as the altar stone. If Stonehenge was a religious temple, as many people believe it was, this may have been the place where the high priests held their ceremonies. All these stones were carefully trimmed into shape by chipping away at the surface with other stones. Larger lumps may have been removed by heating the stones along carefully marked lines, throwing cold water on them, and hitting them. Finally, they reconstructed the bluestones in two groups. Stonehenge was then as complete as it would ever be.

1

The question of how Stonehenge was built seems to be easier to answer than why Stonehenge was built. What was the purpose of this giant stone monument? It used to be thought that it was built by the Druids, who were priests of the Celtic people before the Romans invaded Britain. However, Stonehenge in its final form was completed almost 1000 years before the Druids appeared in Britain.

After years of study Professor Hawkins visited Stonehenge and developed a theory about why Stonehenge was built. He knew that, to the people of the Stone Age, the sun was extremely important. Hawkins reasoned that these people must have considered the sun and moon as gods, for the sun brought warmth and growth and the moon lighted the night.

In 1961 Professor Hawkins stood at the center of Stonehenge and looked through the stone arch leading to the entranceway. He saw the sun on the far horizon rise over the Heel stone! He was excited because his discovery demonstrated that Stonehenge might have been used as a calendar to measure the seasons.

To help prove this, he put charts of Stonehenge into his computer "Oscar." This machine recorded the positions of the different stones, archways, etc. Another machine was used to figure out to what part of the sky each different pair of stones pointed. He concluded that by standing at different places, the builders of Stonehenge could predict when the sun and moon would rise or set in midwinter or midsummer. Through more research, he also theorized that the outermost ring of fifty-six holes was used to predict eclipses of the sun and moon. He found that they take place in fifty-six year patterns. In other words, every fifty-six years an eclipse would take place at exactly that same time of the year. By numbering the holes and placing six stones inside the circle next to certain holes, these stones would be moved from one hole to the next each year. When one of the stones reached a certain hole, the ancient astronomers knew that an eclipse would occur in the spring or the fall of that year.

If all this is true, the people of that time had far more knowledge of mathematics and astronomy than prehistorians ever imagined!

Rock and Roll

Below you will find several definitions for the word *rock*. As you can see, we use the term in many different ways. Select the correct number of definition for each sentence below and then find that number in the code bank. Each number has a corresponding letter. Place that letter at the bottom of the page above the number for that sentence. When you have completed this, you will have spelled out another name for Stonehenge–something that means a stone-standing monument!

rock (rok) n. 1. a large mass of stone forming a hill, cliff, or the like. 2. *Geol.* a mineral matter of variable composition, consolidated or unconsolidated, assembled in masses or considerable quantities in nature, as by action of heat or water. 3. a stone of any size. 4. a firm foundation or support. 5. *Slang.* a diamond. 6. *Slang.* a piece of money. 7. v.i. to move or sway to and fro or from side to side. 8. n. popular music with a strong beat 9. v.t. to affect deeply; to stun. 10. To shake or disturb violently.

1. That's quite a **rock** on her finger!
2. We went to the **rock** concert.
3. Everyone in the courtroom was **rocked** by the verdict.
4. The earthquake **rocked** the entire city.
5. She adopted a pet **rock**.
6. The geologist dug into the **rock** of the mountain.
7. He pulled the **rock** out of his pocket to pay for the sandwich.
8. I really trusted her as a friend. She was my **rock**.

Code Bank	
1 = I	
2 = C	
3 = L	
4 = H	
5 = M	
6 = T	
7 = M	
8 = E	
9 = G	
10 = A	

```
__  __  __  __  __  __  __  __
 1   2   3   4   5   6   7   8
```

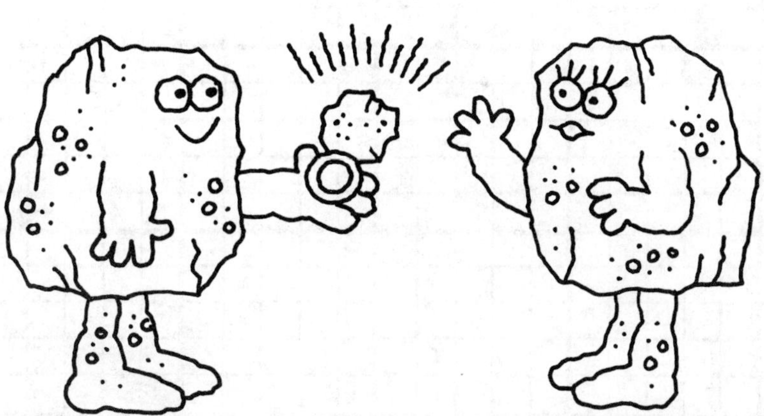

GA1512

And the Plot Thickens!

Usually when you graph two numbers (coordinate points) on a grid, you graph them where two lines intersect. However, with "And the Plot Thickens," given coordinate points, you will graph the entire square by darkening it in. The first number in the pair tells you how far over to go on the grid and the second number tells you how far up to go. The first square (7, 11)–over 7 and up 11–has been done for you. By darkening in the remaining coordinate "squares" you will spell out the answer to the question, "In which age was it the hardest to live?"

(7, 11)	(17, 6)	(17, 10)	(22, 17)	(17, 15)	(22, 12)
(23, 14)	(15, 6)	(19, 12)	(12, 7)	(29, 15)	(17, 8)
(17, 9)	(18, 10)	(17, 7)	(23, 10)	(20, 10)	(11, 8)
(7, 14)	(13, 18)	(20, 8)	(25, 12)	(22, 13)	(24, 20)
(27, 15)	(29, 13)	(17, 17)	(19, 13)	(12, 15)	(19, 15)
(22, 14)	(20, 7)	(13, 14)	(8, 11)	(17, 12)	(20, 6)
(25, 13)	(27, 13)	(19, 19)	(7, 13)	(12, 9)	(9, 13)
(7, 15)	(13, 16)	(15, 7)	(13, 15)	(13, 19)	(13, 10)
(9, 15)	(22, 18)	(11, 6)	(9, 11)	(25, 11)	(13, 13)
(13, 17)	(23, 20)	(23, 6)	(13, 7)	(11, 7)	(29, 11)
(24, 16)	(8, 13)	(9, 12)	(22, 15)	(8, 15)	(24, 13)
(18, 6)	(12, 20)	(28, 11)	(14, 9)	(13, 12)	(19, 17)
(13, 20)	(17, 18)	(13, 11)	(23, 16)	(23, 8)	(19, 10)
(17, 11)	(18, 15)	(19, 11)	(27, 14)	(17, 16)	(19, 18)
(22, 9)	(22, 16)	(14, 20)	(28, 13)	(22, 7)	(19, 8)
(17, 14)	(19, 6)	(24, 8)	(27, 12)	(17, 13)	(25, 14)
(24, 18)	(28, 15)	(23, 18)	(19, 14)	(18, 18)	(15, 8)
(27, 11)	(24, 6)	(18, 11)	(22, 10)	(19, 20)	(17, 20)
(22, 8)	(19, 16)	(22, 6)	(25, 15)	(14, 7)	(22, 19)
(14, 15)	(24, 10)	(22, 20)	(17, 19)	(22, 11)	

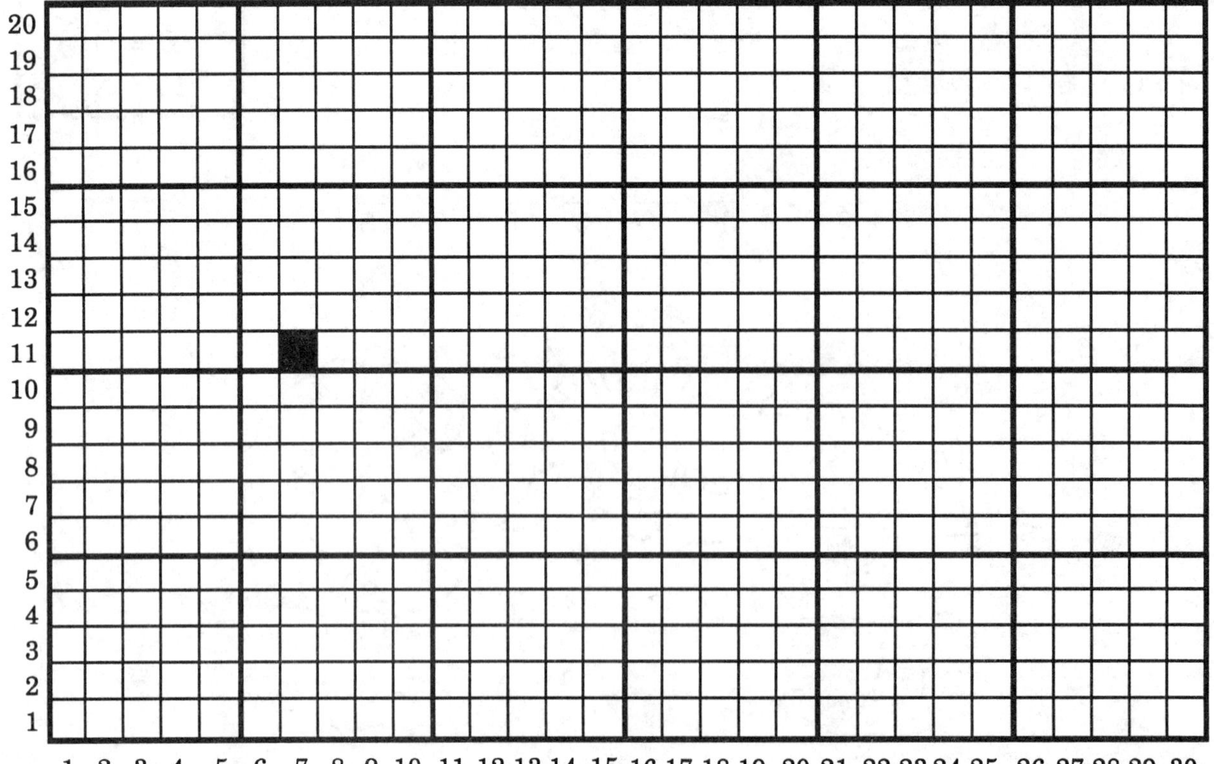

GA1512

Learning Extenders

1. There are several theories concerning how Stonehenge might have been built. Find several of these and argue, in writing, for the one you think is most possible.

2. Make an illustration depicting the construction of Stonehenge.

3. There are many areas of science that might study Stonehenge. They include engineering, architecture, anthropology, archeology, and astronomy, to name a few. Research one or more of these careers (or interview someone in your community) and report how they would be related to Stonehenge. How would people in each field go about studying it?

4. Make a three-dimensional model of Stonehenge using recycled material of your choice.

5. Pretend that you were a newspaper reporter when Stonehenge was built. Write a news story about why it was built, how it was built, and what it looked like at that time.

6. Research Stonehenge and find at least ten questions about this structure that have never been answered.

7. Using number facts about Stonehenge and its construction, write at least ten mathematical problems to share with your classmates. (Use information like the number of different types of stones, the weight of the stones, the number of logs needed to build the structure.)

8. Research the significance of a circle. Why is this shape so important in many religions? Why was Stonehenge built in the shape of a circle?

9. Develop a cartoon strip showing the sequence of the building of Stonehenge. Be sure to include the hauling of the thirty massive pillars, the placement of the lintels, the selection of bluestones, the firing of the stones, the digging of the holes, etc.

10. Research and make a learning center for your classmates on radiocarbon dating, a system used to date objects. Are there other techniques scientists use to tell the age of an object?

You Aren't Getting Any Younger!

An age-old riddle asks the question, "What is it that walks on four feet in the morning, on two at noon, and three at night?" The answer is man. As babies we crawl, then we learn to walk, and in old age we use canes. But the answer to this mysterious question could just as easily be aging; with all the advances in medicine and science, our society is getting older and people are living longer. In 1900 only one out of every 25 people was over the age of 60. Today, it's one in eight and the average life expectancy has increased to 82 years.

Close your eyes and picture an elderly person. What do you see? Someone who has gray hair? Wrinkled skin? Who can't hear you when you talk? Slightly bent walk? We are all very familiar with the signs of aging, but what causes it?

The study of aging is called gerontology. This name comes from the Greek word *geron*, which means "old man." The goal of gerontologists is to discover just how the aging process occurs and how, if possible, to slow it down.

Scientists believe that there are two main factors that affect the aging process. First is genetics. It is believed that our bodies are simply programmed to "wear out" after a given period of time. Take the skin for example. When you smile, your skin wrinkles but when you quit, it goes back to its nice, smoother form–like a rubber band expanding and contracting. In older people, however, that "elastic" has deteriorated because of exposure to the sun and time. The skin holds the wrinkles.

We keep the same brain cells all our lives, but they get fewer as we get older. Aging may simply be due to a loss of cells or structures we cannot renew. Maybe those same chemical changes in the body that make us grow up are the same chemicals that damage the body and make it grow old.

A second theory states that aging is caused by the environment. We grow old as a result of disease and injury outside the body. Germs and bacteria enter the body and over time cause various major organs to no longer function properly or as well. Every part of us is affected from our teeth to our hearts to our immune systems. We cannot escape the toll the environment takes on our bodies, and this may cause us to age.

The chances are about 100 percent that if we are alive we are going to age. But as science continues to investigate why we grow old, it is also experimenting with ways to slow down that process. Some are studying worms and have identified a gene that can increase the life span by almost 50 percent! Others are studying the eating habits of mice. By feeding them a special low-calorie, high-nutrient diet, instead of living 36 months they are living 55 months. The diet has also reduced heart disease and cancers found in these animals.

Still other scientists are studying how to rejuvenate the memory process in the brain to allow some cells to "fill in" when others die.

Is there a "Fountain of Youth"–a way to stop the aging process? Probably not, but there are some things that can be done to "stay younger." Always use a sunscreen when you sun and avoid smoke whenever possible. Exercise regularly–not just your body–but also keep your mind active. Eat right and cut down on the amount of fat you eat.

Growing older is an experience that allows us to enjoy life to its fullest. After all, it's not how long you live, it's how well you live!

6

You Hide Your "Age" Well!

Listed below are the definitions for twenty words that contain the letters *AGE*. First, figure out the mystery word and then locate it in the hidden word find. Words may go vertically, horzontally, forward, and backward.

 1. A small bouquet worn usually at the shoulder
 2. To plunder
 3. A leaf or one side of a leaf as of a book
 4. To hire; employ
 5. A tough, chewy, ring-shaped roll
 6. A barred enclosure for confining animals or birds
 7. An instrument used for measuring or testing
 8. Baggage; suitcases
 9. Distance measured or expressed in miles
10. Violent anger
11. Payment for work done or services rendered
12. A wise person
13. Green fodder that has been stored and fermented in a silo
14. Waste carried off with groundwater
15. Platform for performing
16. The act of drawing or pulling along by a chain or line
17. A cultivation of land
18. A long journey, especially on a ship
19. A rural settlement smaller than a town
20. Electromotive force

```
A G W M S C P B A G E L N
G A S E G A Y O V A W G A
T U G S P I L L A G E M V
V G E G A W O T P E G A B
I E M R G S C O R S A G E
L R A G E M R A L E W E G
L E G A T S E U E W G G A
A G E N G A G E W A S E T
G A G U E G A E L G M A L
E L E G A E L I M E G A O
S L A G G E S R V S W M V
T I E G A C A B I O P G S
V T B A G L G S R C S V A
```

Who's First?

Use the logic clues given for each set of events to see if you can figure out where they go on the time line. Try this first one!

Movie Projector   **Color Television**   **Holography**

Clue A: Thomas Edison (1847-1931) invented the movie projector.

Clue B: Color television was invented after holography.

Where does each invention go on this logic time line?

| 1883 | 1947 | 1950 |

If you thought this through using Clue A you would have placed the movie projector at 1883 because Edison was no longer living at the other two dates. Clue B told you holography was invented before color television. See how you do on the rest of the Who's First logic lines!

Airplane   **Elevator**   **Bicycle**

Clue A: The elevator was invented before the twentieth century.

Clue B: People were riding bicycles before they were riding in an airplane or elevator.

Where does each invention go on this logic time line?

| 1816 | 1852 | 1903 |

Cash Register   **Adding Machine**   **Pocket Calculator**   **Sewing Machine**

Clue A: The cash register was invented before the adding machine.

Clue B: People were able to make their own clothes before stores were able to ring up sales on a cash register.

Clue C: Kilby and Merryman invented the pocket calculator less than 40 years ago.

| 1845 | 1879 | 1888 | 1972 |

Aspirin   **X ray**   **Stethoscope**   **Hypodermic Syringe**

Clue A: Doctors could prescribe aspirin for headaches before they could X-ray heads.

Clue B: Doctors have been able to listen for a heartbeat for over 200 years.

Clue C: Doctors could give shots with syringes before they could prescribe aspirin.

| 1781 | 1853 | 1889 | 1895 |

Learning Extenders

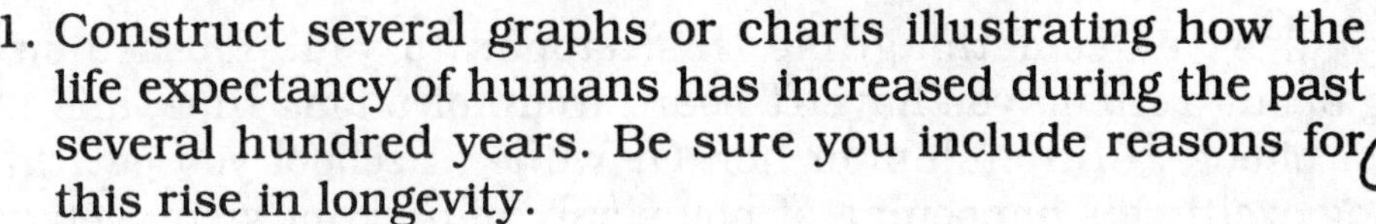

1. Construct several graphs or charts illustrating how the life expectancy of humans has increased during the past several hundred years. Be sure you include reasons for this rise in longevity.

2. Spanish explorer Juan Ponce de León searched for the "Fountain of Youth." Research Ponce de León and make a shadow box showing his "discovery" of the fountain.

3. Progeria is the disease of premature aging. Research this topic and develop a knowledge center for your class.

4. Produce a "This Is My Life" coin collection. First, locate a penny for each of the years since you were born. Glue the pennies on a piece of cardboard and next to each write:
   a. an important personal event that happened during that year
   b. an important national or international event.

5. Develop a list of questions that you would use when you interview a retiree in your community. Have other students in your class interview retirees they know and work to develop a "Senior Citizen" center in your classroom.

6. Research Alzheimer's disease and create an outline illustrating your findings.

7. Write a story or poem titled "The Boy (or Girl) Who Never Got Old."

8. Write to several agencies or organizations on aging. (The National Institute of Mental Health may be one.) Collect information and put together a class learning center on aging.

9. Identify fifteen key words or terms related to the topic of growing old and write their definitions. See how many of these your classmates know.

10. Research several "drugs" that have been found to slow down the aging process and report to your class in the form of a video magazine or newspaper. You might want to think about Retin A (helps "remove" wrinkles by stimulating cell growth) or Vitamin C (which may help make collagen).

11. Different countries of the world treat older people in different ways. Select five countries and research their treatment, and then rank those five in the order you feel is most appropriate and humane.

GA1512

ESP

Maybe you've had something like this happen to you. You are sitting at home thinking about a friend you haven't heard from for a long time, and all of a sudden you get a phone call from him or her. Or while at school you get this feeling that something exciting is happening at home only to find out when you get there, your parents have bought you a new pet! If any of this sounds familiar, you're among millions of people who have experienced ESP.

Extrasensory perception, or ESP, is part of the science called parapsychology. It was first described over one hundred years ago by Dr. J.B. Rhine as a way of getting information without using any of our five senses—hearing, seeing, smelling, touching, tasting. In fact, some say that ESP comes from our "sixth sense." Dr. Rhine thought everyone had some ESP. You can decide if you have a little or a lot!

Clairvoyance

Clairvoyance is the ability to "see" things which are not visible that are going on at that time. Sometimes this is called a "vision." An example of this would be someone who "gets the feeling" that his or her house is on fire only to find out later that a pan of grease did catch fire in the kitchen while dinner was being prepared.

You can see how clairvoyant you are by doing this experiment. Take a deck of cards. Have someone shuffle the deck and remove a card. Place the card facedown without either of you looking at it. Clairvoyance is the ability to know what that card is.

Telepathy

Telepathy is the ability to "read someone's mind." You might have heard someone say, "I knew that was what you were thinking." Sometimes you probably think that your parents can read your mind!

With this test you can see what telepathic powers you have. Find a partner and a deck of cards. Shuffle the cards. Have your partner take one card at a time and look at it. That person should think about the card he or she is looking at and try to send you a message as to the suit of that card. For example, if it is the king of diamonds, the person would think diamonds. Your ability to tell that card is a diamond is telepathy.

Precognition

Precognition is the knowledge of events before they happen. In 1858 Mark Twain and his brother Henry were working on separate boats. One night Mark had a dream where he saw his brother in a metal coffin, dressed in one of his suits, with a bouquet of flowers with one rose in it lying on his chest. A few days later Mark's brother's boat exploded and he was killed. When he arrived he saw his brother in a metal coffin, dressed in one of his suits, with a bouquet of flowers with one rose in it lying on his chest.

Do you have precogniton? Get the cards back out. You tell your partner that the eighth card is the king of diamonds. Your friend shuffles the cards a few times, counts to eight, and turns the card over. If it is the king of diamonds, you have precognition.

All of this sounds pretty strange, and many scientists are more than skeptical about this new sense. Then again, many scientific theories were considered pretty strange when they were first introduced. Maybe someday we'll know the truth about ESP.

How Did You Do That?

Have you ever wondered what it would be like to read someone's mind? Some people say they can, but most of the time it's just a trick.

Here are some "mind reading" tricks you can pull on your family and friends but be careful, even *you* may think you have ESP.

The Disappearing Number

Have someone blindfold you. Ask that person to write down any four-digit number like 2345 or 5478. Any four-digit number will do. Ask him or her to multiply the number by nine.

Then tell him or her to select any digit of this large number except 0 and erase it. Tell your friend that you will give him or her the number he erased if he or she will tell you the sum of the remaining digits.

How do you do that? Let's try it. Let's say I pick the number 5478. When I multiply it by 9, I get 49,302. I choose to erase the 3. If I add the other digits together, I get 15.

You could tell me which number I erased by subtracting the sum of the remaining digits (15) from the nearest multiple of 9 which is higher than the sum in question (18). 15 from 18 is 3, the number I erased!

Your Answer Is Nine!

Here's a way "to force" your friends to think of a problem where the answer is always nine.

Put into your calculator any number from 10 to 99. Reverse the digits. Then subtract the smaller from the larger one. Divide by the difference in your original two digits. The answer will always be nine.

Let's try it. Say I pick 41. Reversing the digits, I get 14. Subtracting 14 from 41 I get 27. Divide 27 by the difference in my original number which is 3. My answer is nine!

Brother, Can You Spare a Dime?

Here's a "trick" that will allow you to tell a person not only his or her age but also how much loose change is in his or her pocket. Sounds strange? Just try it!

Give a friend a piece of paper to use for this activity. Have him or her write his or her age on a piece of paper without showing it to you. Now tell him or her to multiply it by 2 and then add 5 to this answer. Multiply that answer by 50. Subtract the number of days in a year, 365 (not a leap year).

Now have this person count the amount of change in one pocket or purse that is under one dollar. Add this number to the total so far. Have the person give you the final answer and you can tell the age and amount of loose change. How? Let's try it!

Suppose I am 41. So I would multiply this by 2 and get 82. Add 5, which equals 87. 87 multiplied by 50 is 4350. Subtract 365 from 4350 and get 3985. Let's say the change in my pocket is 47 cents. 3985 plus 47 is 4032. This is the number I would tell you.

Now quickly add 115 and your number to work with is 4147. The first two numbers tell my age and the second two give the amount of change in my pocket! What a trick!

ESP and Other Abbrev.

Our lives are full of time-saving devices—microwaves, portable telephones, and fax machines just to name a few. Sometimes even the way we communicate is shortened. For example, you might say, "Mom, may I go to play b. ball?" She may or may not know you're headed for the basketball court. Or you might say, "Dad, can we watch MTV instead of PBS?" Below you will find some scientific abbreviations. The letter next to the correct answer will spell out something you are about to discover about yourself.

1. R.E.M. H. Rapid Eye Movement Y. Rover Extravehicular Module E. Radar Emitting Monitor

2. H.I.V. X. High Intensity Volume O. Human Immunodeficiency Virus M. Hologram in View

3. D.N.A. W. Deoxyribonucleic Acid T. Dietary Nutrition Assistance L. Department of National Assessment

4. A.C. S. Assisting Computer P. Alternating Current R. Access Charge

5. L.A.S.E.R. E. Light Amplification by Stimulated Emission of Radiation T. Last Action System Engage Response D. Liftoff Astronaut Sensor Emergency Reaction

6. F.M. P. Fault Mountain C. Formulation Method R. Frequency Modulation

7. P.M. F. Previous Maintenance C. Post Meridian W. Polar Mode

8. R.O.M. E. Read-Only Memory H. Repaired Organ Monitor O. Resonance Oscillator Machine

9. S.E.E.R. P. Seasonal Energy Efficiency Rating B. Space Environmental Engineer Robot D. Science Environmental Education Research

10. N.A.S.A. T. National Aeronautics and Space Administration G. Natural Acoustics and Sound Amplification M. Neutral Acidic Sensing Apparatus

11. C.A.D. F. Cosmic Astrological Detector H. Cardiac Arrest Device I. Computer-Aided Design

12. D.A.T. I. Distinct Arterial Trauma V. Digital Audio Tape D. Direct Air Transportation

13. F.D.A. E. Food and Drug Administration K. Federal Dietary Association G. Foreign Developed Application

14. E.P.A. N. Empirical Pasteurization Activity U. Environmental Protection Agency B. Epidemic Patron Assistance

15. C.R.T. G. Chemical Response Testing R. Cathode-Ray Tube O. Cosmic Reaction Theory

___ ___ ___ ___ ___ ___ ___ ___ ___ ___ ___ ___ ___ ___ ___!
 1   2   3   4   5   6   7   8   9   10 11 12 13 14 15

Learning Extenders

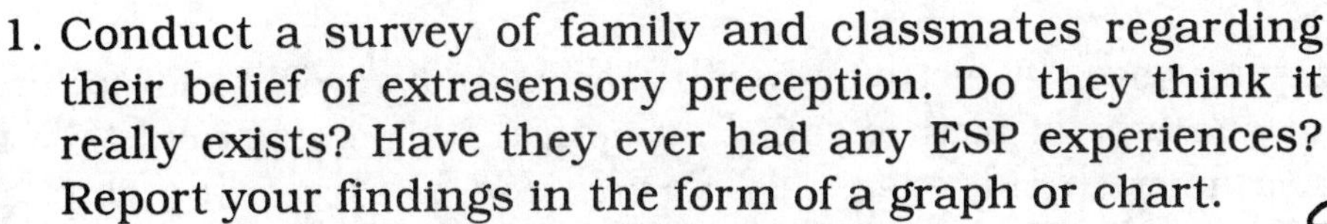

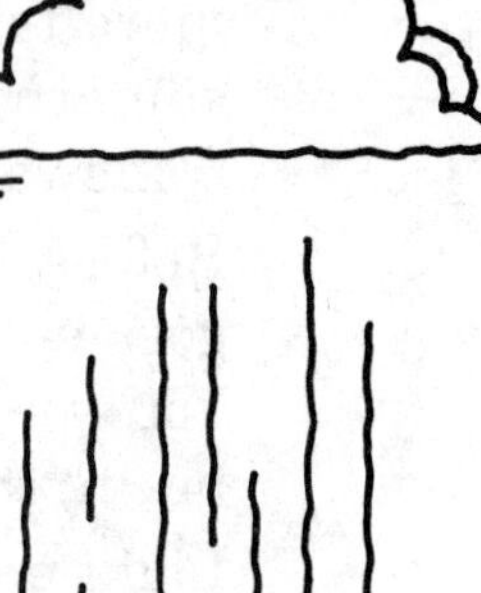

1. Conduct a survey of family and classmates regarding their belief of extrasensory preception. Do they think it really exists? Have they ever had any ESP experiences? Report your findings in the form of a graph or chart.

2. Locate a book on ESP in your school or public library. Read the book and write an evaluation of it. Would you recommend it to anyone else? Give five reasons for your choice.

3. Find out more about famous ESP events (like Mark Twain's). Put them into the correct category as being clairvoyance, telepathy, or precognition.

4. Compose experiments that would confirm or deny someone's claim of having ESP. A good place to start is with a book titled *ESP: Your Psychic Powers and How to Test Them* by W.R. Akins.

5. Compose a song or poem about a famous ESP event.

6. Pretend you are a newspaper reporter sent to cover a famous ESP event. Write the story that you want to appear in the newspaper.

7. Produce a "Believe It or Not" coloring book on famous ESP events.

8. Interview someone in your community with knowledge on the subject of extrasensory perception. Report your findings back to your classmates.

9. There are several major theories about how ESP actually works. One says it is a leftover talent from before language was developed. Another says it has to do with time. We think of time only going forward. But what if it goes backward or sideways every once in a while? Evaluate your findings and report the "most probable" to your class.

10. Set up a center where classmates can test their ESP.

Optical Illusions

A famous saying goes, "A picture's worth a thousand words," and nothing could be more true of one type of picture called an optical illusion. *Optical* tells that it has something to do with the eyes, and an *illusion* is something that appears in someway different than it actually is.

Before you can really understand these illusions, you must first understand what goes on behind open eyes. Keep in mind that it's not just your eyes that let you see. You see because of the cooperation of nerve cells in your eyes, nerve cells in your brain, the optic nerve, and light. Briefly, this is how it all comes together.

Light enters your eye through a clear cover called the cornea, which bends the light to pass through the iris. The iris controls the amount of light coming into your eye through a hole called the pupil–the colored part of the eye. Muscles in the pupil allow it to open wider to allow more light to enter and get smaller to allow in less light.

Just behind the iris is a flexible lens that focuses the light onto the retina. The retina contains more than one hundred thirty million sensitive nerve cells. There are two kinds of nerve cells located here: rods and cones. The cones let you see colors and sharp detail. They work well in bright light, but not in dim light. The rods let you see only in black, white, and gray. They work only when the light gets really dim. When light strikes the retina, nerve cells respond and send the brain electric currents called nerve impulses.

The impulses travel along cells to the optic nerve and from there your brain's visual center. It is here in the brain that the information as an impulse is translated into a language your brain can "read" and interpret. The two eyes send slightly different signals which allow us to see in three dimensions. Remember, seeing calls for some thought and some experience. It is the brain that makes sense of the image information entering the eyes. Because of past experiences, our eyes are accustomed to seeing things a certain way and expect to see it that way every time. So what does all this have to do with optical illusions, you ask?

Remember, to see anything requires more than your eyes. If you were reading this at night and the lights went out, you couldn't see to continue. Light, your eyes, and your brain work together to enable you to see. If any one of these three is missing, you cannot see. And if any one of them is "confused," an optical illusion is the result. An illusion fools the brain because the brain has learned to expect to see a picture a certain way and tries to make it into something it has seen before.

Illusions provide us with mysterious pictures that produce errors in perception and cause us to develop a false impression of the "facts" presented to our senses. Lines may seem longer than they really are, straight lines may appear to be curved, colors may be there that aren't, or a two-dimensional picture may actually look 3-D. Whatever the case, optical illusions create an interesting and exciting challenge for our "mind's eye."

Where's Fido?

The problem: Fido is not in his doghouse. How can you get him there?

The answer: With a thaumatrope!

The word *thaumatrope* comes from two Greek words meaning wonder and to turn. A thaumatrope works because, when an impression is made on the retina of the eye, it lasts for about one-eighth of a second after it is taken away (called retinal retention). It's because of that fact that when we watch television or go to the movie theater our eyes make us think we're seeing "moving pictures" when actually it's just a series of still pictures moving super fast!

By following the instructions below, you can get Fido back into his doghouse and then make a thaumatrope of your own.

1. Cut out the two squares of Fido and his doghouse and glue to a piece of cardboard cut the same size. (Be sure to check and make sure Fido won't be upside down when you "flip out.")
2. Punch a hole near each of two opposite edges.
3. Attach a piece of yarn or string as shown.
4. By sliding the thumbs over your fingers, you can twirl the square rapidly and presto, Fido's back home!

15

Now You See It . . . Now You Don't

One of the interesting things about optical illusions is that not everyone sees things the same way. Use the "test" below to find out how your friends see. Keep a record of the data you collect.

1. Does this drawing go down like a tunnel or up like a projection?

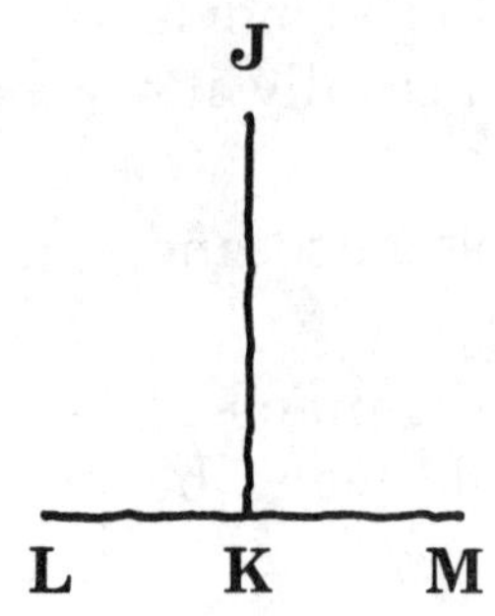

2. Carefully observe the picture on the left. Is the vertical line JK longer, shorter, or equal to the horizontal line LM?

3. What do you see in this picture?

4. Do you see a pretty young girl or an old woman?

5. Is this book turned in or turned out?

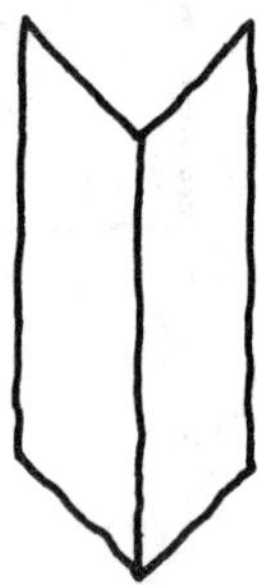

6. Is this hat taller than it is wide?

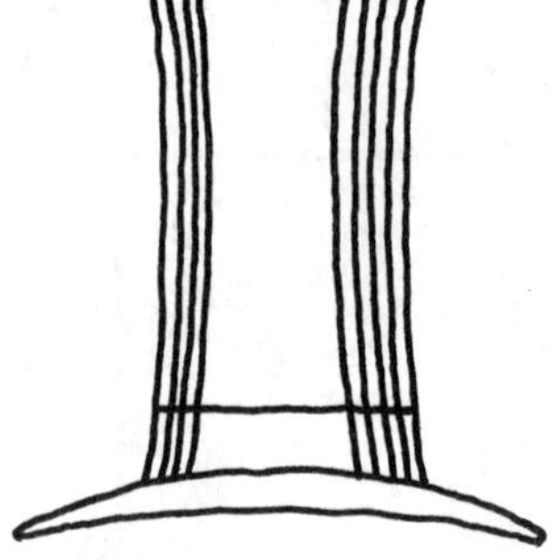

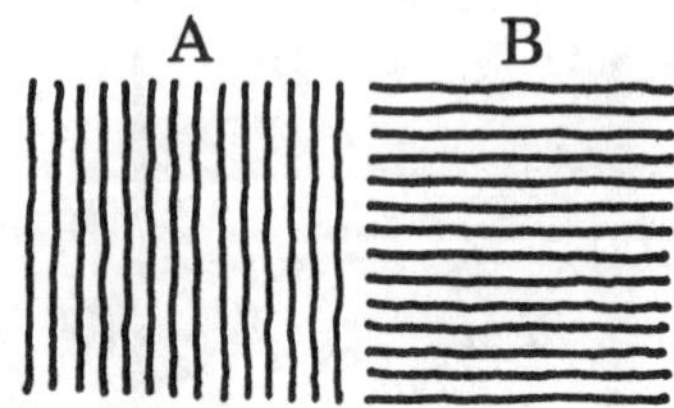

7. Is drawing A as high as it is wide? Is drawing B as wide as it is high?

16

Learning Extenders

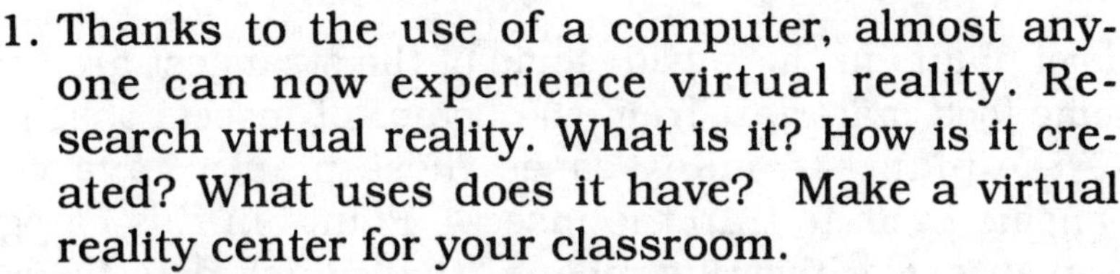

1. Thanks to the use of a computer, almost anyone can now experience virtual reality. Research virtual reality. What is it? How is it created? What uses does it have? Make a virtual reality center for your classroom.

2. To demonstrate how a cartoon fools you into thinking you see a smooth motion, construct a phenakistascope. Directions can be found in the February 1993 edition of Scholastic's Super Science (blue).

3. You can create your own "motion picture show" with a flip book. Simply take fifteen to twenty small pieces of paper the same size and draw a picture on each making sure that you change each picture just a little. Staple all the pages at the top and then "flip it" to get motion.

4. Create a class learning center on optical illusion activities that use one eye, that involve seeing colors that aren't really there, and that show movement of objects. Place the activities in a center and have the class share their experiences at the end of the week.

5. Some people who choose careers in motion pictures are very concerned that they are able to create images that really aren't there. These people are called FX (short for effects). If you were to "direct" a new movie you have created, what special effects would you want it to have? How would you go about creating those effects? Design a movie poster advertising your movie.

6. Now that you know about optical illusions and how they work, design some of your own. Convince other classmates to do the same and have a class contest for the "Best New Optical Illusion."

7. People who are visually impaired face many challenges in today's world. Locate a biography of a visually impaired person and create an original book jacket for it that gives a short summary on the inside of the jacket.

8. Our eyes are amazing organs. Using recycled materials, construct a model of the eye that can be taken apart and put back together, and share it with your classmates.

Carnivorous Plants

Of the thousands of plants in the world, some of the strangest are the ones that are able to obtain some food materials from the bodies of insects that get caught in or on some part of the plant. In many cases these plants have very specialized structures that enable them to trap the insects. Plants in this category are known as **carnivorous** plants. Carnivorous means "that which eats flesh." These words can be misleading. While these plants do "eat" insects, they are not able to behave like animals and actively catch and eat them. In every case, the insects are attracted by the plants' scent or color and may become trapped there. In time, parts of the insects are digested and absorbed by the plants.

There are about 500 different kinds of carnivorous plants throughout the world. In spite of all the stories about huge plants that eat humans, none of this is true. There is, however, some truth to reports of small rodents falling into pitcher plants and not being able to escape. In time, their bodies are digested.

Even though stories about man-eating plants are not true, carnivorous plants are still very interesting. These insect-eating plants grow in bogs, in damp sand, or along the distant edges of lakes and rivers. These plants are fragile and cannot stand changes in the conditions around them. If any pollution should seep into the bog where they are living, they will be some of the first plants to die.

All carnivorous plants are very similar to other green plants. They have leaves and stems, but they lack a large root system. Their leaves are green, they have chlorophyll, and they are able to obtain energy from the sun. All carnivorous plants can exist without eating insects. However, where the soil is deficient in certain elements that the plants need to grow, particularly nitrogen, the plants are able to obtain these minerals from digesting the bodies of small animals and probably grow better than carnivorous plants that do not have insects as a food supplement.

The shape of the leaves of carnivorous plants differs from ordinary plants. Where the leaf of an ordinary green-leaf plant is usually flat, the leaves of insect-eating plants are often shaped in some special way to catch or hold insects. In both cases, leaves take energy from the sunlight, but in carnivorous plants these leaves may also be able to trap food.

Insect-eating plants differ in the way they catch or hold their prey. The traps on carnivorous plants can be divided into three general groups.

Plants with pitcher or bladder-like leaves. The pitcher of the pitcher plant is really a very specialized leaf, and it contains a pool of liquid into which insects fall and are digested. To make it difficult to get out of the leaf, there are slippery hairs that point downward. The bladderwort is an underwater plant with bladders. Each bladder has a small opening at its bottom surrounded by bristles. Tiny insects swim along the bristles and when they touch one, they are sucked into the plant.

Hinged leaves. The end of each leaf of a Venus's-flytrap has hinged halves that can snap shut like a book. Each leaf-half has trigger hairs. Insects landing on them spring the trigger and the leaf snaps shut, trapping the insect. There are long, thin teeth on each half and they keep the insect from getting out.

Sticky leaves. Sundews' round leaves are covered with hairy glands that make a gluey liquid. Insects that land on a leaf are stuck fast. Surrounding hairs on the plant then curl around the insect and acid juices digest it.

Finding Your Roots!

All over the world, people are tracking down information about their "roots" as part of a fast-growing hobby–genealogy. Even before there was writing, people were keeping track of their family history by memorizing information and passing it down from generation to generation.

The best place to start your detective work is with your parents. Ask them to tell you stories they can remember about you and other family members. Be sure to write down the information or tape record it! Next, talk to your grandparents and great-grandparents if you can. If they are no longer living, talk with older members of your family who can give you the information you need. Try to collect pictures or newspaper clippings from the past and make a family scrapbook.

Some of the questions you might want to ask could include:

1. What is your full name?
2. Were you named after someone? Who?
3. Where and when were you born?
4. Where did you grow up? What schools did you attend?
5. Where did you meet your spouse?
6. Where and when did you get married?
7. What did you and your spouse do for a living?
8. Where did you live after you were married? Did you move around?
9. What hobbies and interests did you and your spouse have?
10. What special family traditions did you celebrate?
11. What do you know about the history of the family name?

Of course, there are many other questions. Ask those that interest you but remember, write down everything. You may not think it is important at the time, but it may be later.

When you've gathered all the information you want, fill out a pedigree chart. A pedigree chart, like the one below, outlines your family's "roots" at a glance. You probably can't put all the information you collect on it, but you can put some information about each family member. To begin your tree, start with yourself on the far left.

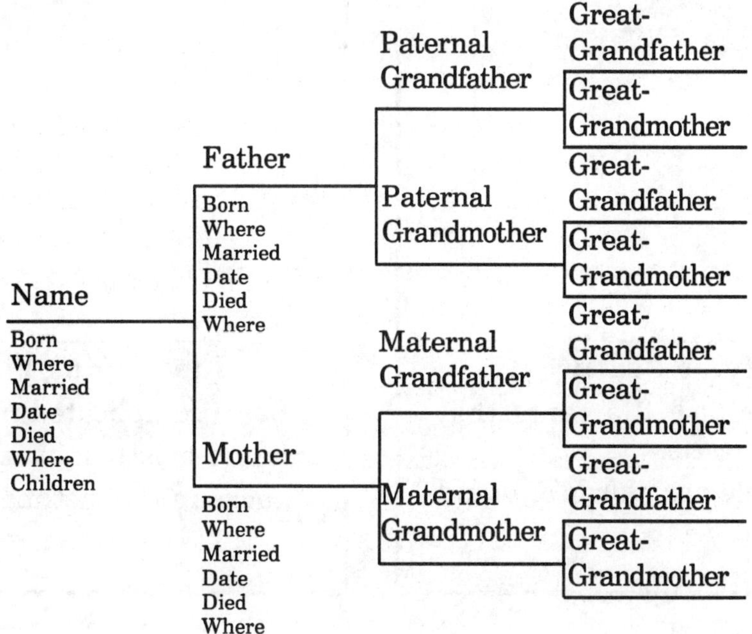

Wanted: "Man-Eating" Plants

The famous "Man-Eating" Plant Gang has just escaped, and it is your job to get the wanted posters out as quickly as possible. Using the information given, draw each plant on the correct poster.

Wanted

Name: *Sarracenia*

Alias: ______________

Identifying Features

√ has a tube or funnel topped with a hood

√ has a smooth, waxy inside

Wanted

Name: *Dionaea*

Alias: ______________

Identifying Features

√ looks like a clamshell

√ has "trigger" hairs

Wanted

Name: *Drosera*

Alias: ______________

Identifying Features

√ has 200 reddish filaments that look like tentacles

√ each tentacle has a drop of liquid on its tip

Wanted

Name: *Utricularia*

Alias: ______________

Identifying Features

√ has a complicated trapdoor

√ found mostly in watery areas

√ sucks in the victims

Learning Extenders

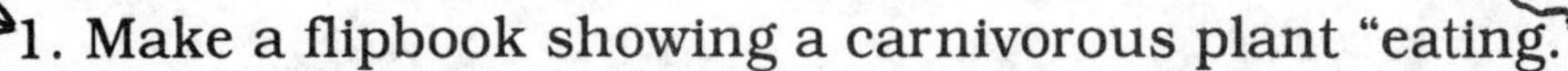

1. Make a flipbook showing a carnivorous plant "eating."

2. On a world map, locate the various locations where carnivorous plants can be found.

3. Pretend that you are a botanist and are trying to develop a new species of "man-eating" plant. What would it look like, where would it grow, what would it eat, what would you name it?

4. Construct a game using information about carnivorous plants that can be shared with your classmates.

5. Contact a local garden center and interview an employee on the subject of carnivorous plants. Find out if there are any in stock. (Many greenhouses do have Venus's-flytraps because they can exist in many environments.) See if you can purchase a pet plant for your room.

6. Write a story titled "A Day in the Life of a Sundew." Make sure you write your story as if you were the plant. Tell all the things that might happen to you in one day.

7. You have just been asked by a local museum to construct a display on carnivorous plants. Using only recycled materials, construct that flower garden display. Make sure you include facts about each flower.

8. Determine the five most important facts you learned about carnivorous plants during your reading. Rank them from most important to least important, giving reasons for your choices. Compare your ranking with other classmates' rankings.

9. Pretend that you are an attorney, and a client named Venus Sundew has just come to your office to hire you to defend him on the charge of "man-eating." With the assistance of one or more classmates (who could serve as prosecutor and judge), hold a mock trial with the remainder of your class being the jury.

10. Locate a book on carnivorous plants from your school or local library. After reading it, list three or more things that you would still like to know about carnivorous plants that were not covered in the book.

11. You are going on vacation and have asked your neighbor to take care of Munchie, your man-eating plant, while you are gone. However, Munchie is very particular. Write a handbook on the care and feeding of carnivorous plants.

Weather Forecasting . . .
Listening to Nature

Have you ever made special outdoor plans for the weekend only to have it rain and spoil the whole thing? Most likely you have. Wouldn't it be great if you could correctly forecast the weather? Believe it or not, you can without using the television, radio, or newspaper. If you just listen to nature, you, too, can become a meteorologist. Of course you have to remember that even ir the "real world" of professional forecasting, nothing is 100 percent accurate.

Thanks to the advancements in science, weather people can use satellites in space to collect a variety of information and relay that back to earth. Additionally, meteorologists use a type of weather radar called Doppler. Doppler radar detects precipitation and wind circulation within a cloud. By seeing the wind pattern within a storm, forecasters may be able to tell if a tornado is developing. Doppler radar calculates wind by detecting particles moving either toward or away from the radar. The echo from a particle moving toward the radar is different than an echo from one moving away. The radar determines this and indicates how the particle is moving.

Computers also play a major role in weather forecasting. Computers help to quickly collect data, calculate how a pattern might change, and provide maps that show the type of weather that can be expected. From all this, short-term predictions are usually quite accurate. But when you get past 24 hours, the accuracy of the forecast is not as reliable.

You probably don't have a high-speed computer, massive satellites, and expensive radar at your disposal. However, you, too, can predict weather by using a few rules and some nature observations.

Pressure

The weight of air pressing on the earth is called air pressure. Cool air produces more pressure than warm air, so a rise in pressure often means that weather will be cooler and drier. Falling pressure indicates warmer, wetter weather.

Wind Direction

Wind direction has an effect on temperature. By studying the wind around a low and high pressure area, it has been found that wind around a low pressure system blows counterclockwise and means you should expect a warm southerly wind ahead of a low and a colder northerly wind behind it.

Cloud Cover

Different kinds of clouds bring different kinds of weather. Cirrus clouds are always very high in the sky and look feathery. They are made of tiny drops of ice and are often a sign of weather change. Stratus clouds usually follow cirrus clouds and can cover the entire sky, making it very gray. They often produce light rain and drizzle or snow. If stratus clouds become dark, heavy clouds, they are called nimbus. These clouds are usually low in the sky and bring heavy rain or snow. The cumulus clouds are puffy-looking, white clouds that usually accompany fair weather.

Looking to Nature

A long time ago, when people lived mostly out-of-doors, they were very close to nature. They soon noticed that plants, animals, insects, and birds sensed the coming of a storm sooner than people did. Using nature's information, forecasts could be made helping people to stay dry and helping farmers to know when to plant and harvest their crops.

Like these early people, you, too, can find many weather signs among wildlife, because of their highly developed senses. For example, take mice who, sensing a drop in pressure, will run out of their holes and squeak and frolic before a storm. A drop in air pressure probably produces this behavior, but it might also be caused by infrasounds. These sounds, too low for human hearing, are given off prior to many natural disasters–tornadoes, severe storms, earthquakes, etc.

Birds and bats also have very sensitive ears and can feel the change in pressure. They show this change in many ways. Some become irritable and quarrelsome and will fight over a piece of food. Some will sing very loudly. If you are a bird-watcher, you know that they feed early in the morning. However, if a storm is coming, they will feed late in the day. If you see birds flying in an unusual or erratic way–soaring and dipping with their wings almost touching the ground–this usually means unstable air or a change in conditions which brings showers.

If you live around a sandy beach, watch for it to come alive with sea crabs. If it does, it's time to head for home. A change in air pressure over the water warns the crabs that a storm is coming. Watching squirrels can also indicate rain. They may be extra frisky, playing follow the leader and "fighting" among one another. All of a sudden they stop and begin patching up and reinforcing their nests. You can be sure of a downpour soon.

Many other animals also predict rainfall. Spiders actively work in their webs before a rainfall. Opossums and raccoons carry their young away from their hollow log homes just before a flooding. Deer leave the high ground and come down from the mountain before a storm. Many small insects will attempt to move into the house right before a rain. Those that do not can be seen busily building up mounds of dirt around nest holes. About two hours before a rainfall, they will quit and go to their nests until after the rain.

The cricket and katydid can help you determine how hot it is outdoors. The cricket is often called the "poor man's thermometer" because it is said to chirp more frequently as the temperature rises. To compute the temperature in the summer in Fahrenheit, add thirty-seven to the number of cricket chirps in fifteen seconds. For katydids, add fifty-four to the number of calls you hear in twenty seconds.

No matter where you live, you will find many kinds of creatures that can help you predict the future weather. Ants, fleas, roaches, flies, crickets, cats, dogs, mice, rabbits, chipmunks, and a variety of birds will all provide you with the information you need. If you observe them closely, using a little bit of information, a little bit of observation, and a whole lot of luck, you, too, may become an ace meteorologist. Why don't you give it a try!

Picture This!

Before there were weather people or weather reports in newspapers, people depended on signs in nature to forecast the weather. Farmers watched the skies before a storm, travelers watched the leaves blowing in the wind. Sailors watched the clouds changing. After a while people began making up rhymes about the weather to help them remember what those signs meant. These rhymes were so popular that they were printed in almanacs. Many of these sayings are still in use today.

Select one or more of the following weather sayings to illustrate, or find some of your own. You might want to make a book of your drawings to share with your classmates.

When clouds look like rocks and towers, the earth will have many showers.

The higher the clouds, the better the weather.

Wooly sheep in the sky will bring raindrops by and by.

A sunny shower won't last half an hour.

Evening red and morning gray send the traveler on his way.

Evening gray and morning red send the traveler wet to bed.

Sea gull, sea gull, sitting in the sand; it's never good weather when you're on land.

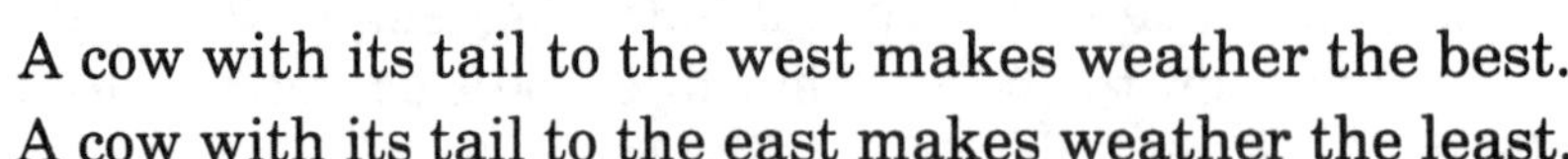

A cow with its tail to the west makes weather the best.
A cow with its tail to the east makes weather the least.

When the wind is from the east, it's not good for man or beast.

Rainbow in morning, sailor takes warning.
Red sky at night, sailor's delight.

Rain before seven, clear before eleven.

Write Now!

There are many things you can find out about meteorology and weather forecasting. Select an organization, business, or individual to whom you could write for information about a career related to weather. Follow the steps for effective business letter writing in preparing correspondence to them.

Write An Effective Business Letter

1. Plan

- Think of your purpose. Decide what you are going to write about.
- Do some research so that you will be knowledgeable about the situation. Decide on what information the receiver must know to respond to your letter.
- Decide who is the appropriate person to write. Be sure that the person or company is able to do something about your request for material or information.
- Outline the letter. Jot down the points you want to include.

2. Write

- Include the six parts of a business letter:
- Heading: Always include your full name and address in a business letter.
- Inside address: Include the name and full address of the person or company who will receive the letter. Be sure to use the person's complete title if known.
- Salutation: If you do not know the name of the person who will receive the letter, use Dear Sir or Dear Madam.
- Body: In the first paragraph, be sure to identify yourself and the reasons you are writing the letter. Make it clear why you are interested in having a response. Clearly state what you want the person to do. The body of your letter should be reasonably short but complete. Make sure you are polite in your letter, and conclude by thanking the recipient for reading it.
- Closing: Use a formal expression like Yours truly or Sincerely.
- Signature: Be sure to write your full name clearly, since this is the only place your name will appear on the inside of the letter.

3. Proofread and Revise

- Reread the letter carefully and make sure you have covered all the points you listed in your outline. Check for any spelling, grammatical, or punctuation errors. It's a good idea to ask someone else to read it in order to double-check your efforts.

4. Rewrite

- After making all the necessary corrections, rewrite the final copy of your letter in ink or type it. You're now ready to put it in the mail.

GA1512

Learning Extenders

1. Interview a grandparent or family elder as to weather sayings they remember their parents following. Are these still believed today? You might want to tape record your interview and share with your class.

2. It takes special training to be a meteorologist. Make a list of the subjects you study in school and tell why knowledge in each of those areas is important to a meteorologist.

3. Many other animals (than the ones mentioned in the information page) are said to be able to "forecast" weather. Make a list of ten animals that can be observed and tell what their behavior means in weather forecasting.

4. Research the various types of clouds and complete a chart telling what each type means to weather forecasters. Keep in mind that the wind direction makes a difference in what the cloud formations mean!

5. Construct a weather board game that teaches classmates interesting and important facts about meteorology and forecasting. Be sure to include signs in nature!

6. Keep a weather chart for one month. Be sure to record date, time, temperature, barometric pressure, relative humidity, wind direction and speed, cloud type, and weather description. From this information, try to predict the next day's weather and keep track of how accurate you are.

7. Reading a weather map can be very difficult. Research and learn about the symbols used on a weather map and then write a short children's book explaining how it is done.

8. Throughout history we have tried to change our weather. If it's sunny, we want it to rain. If it's raining, we want sun. Many American Indian tribes believed that through dances they could affect the weather. Read about their attempts to change the weather and create a "Sun Dances, Rain Dances" center for your class.

9. There are many violent hurricanes and tornadoes that "hit" the United States. Locate a reading source on how meterologists predict these storms. What instruments do they use? How do they know a tornado is about to happen? Answer these and any other interesting questions you can think of.

10. Investigate the function of the Beaufort scale, and make a chart to show your findings. Record the wind speed in your community for one month, and record where on the scale it would fall.

11. Make a list of twenty occupations. Include jobs like wheat farmer, offshore fisher, restaurant owner, ambulance driver, etc. Tell how weather might affect each occupation.

Mysteries of the Mind: Using Your Memory

Think back to the time you were very, very young. What is the first thing you can remember? Saying *Mama* or *Dada*? Crawling or walking on your own? Some other significant event in your life? Whatever you can remember is information that you have managed to store in your memory.

Can you imagine not being able to recall anything? All your experiences would be lost as soon as they ended, and you wouldn't really *learn* anything.

Scientists know little about what happens in the brain when memories are stored. However, they are almost certain that storing new memories involves both chemical changes in the nerve cells (an increase in the production of RNA, a chemical that makes proteins) and changes in the nerve cells' physical structures. This all happens in a small part of the cerebral cortex of the brain.

There are three different stages of memory. The first is **immediate** or sometimes called **sensory** memory. This holds information coming in from the senses for only a short period of time, less than a second. For example, you look at a picture of a sandy beach. The picture passes from your eyes to your sensory memory, which briefly holds a nearly exact image of the picture for less than a second. For this information to last, it must be transferred to a second stage of memory–**short-term** memory.

Your short-term memory can hold five to seven facts for as long as you are able to think about them–usually about thirty seconds. For example, you look up a phone number in the directory. You can usually remember the number until you dial it and then often forget it. Unless you continually repeat the number, it doesn't stay in your head very long. Let's say that it is an important number that you want to remember. Then what happens to it? Well, it goes into the next stage of memory–**long-term** memory.

This memory holds a huge amount of information, some of which will last a lifetime. Information gets to this stage one of two ways. Either you repeat it over and over again or it's put there through some intense emotional situation, like a car wreck or falling in love. If you know your multiplication facts, they're in your long-term memory.

Is it necessary to remember everything? Probably not. Much of the information that enters our brain during the day needs only to stay in the sensory memory stage. It's all right to forget some things after they happen. For example, if your mom or dad sends you to the store to get milk and bread, it's only important that you remember that information until you get to the store. It's not something you need to put into your long-term file.

If you have trouble remembering things, you may be in luck. Memory experts believe that with practice, you can improve your memory. You might try using a mnemonic device (see page 29) or associate the information with a visual image or categorize the information or link it to something else. There are many brain exercises you could practice that might make you a modern day Themistocles, the Greek statesman who is said to have known the names and faces of 20,000 Athenians!

Memory Overload

Here's a chance to challenge your sensory memory (and maybe your short-term memory too). Cover the questions at the bottom of the page and study the picture at the top of the page for two minutes. Then cover the picture and uncover the questions. See how many of them you can answer in two minutes.

1. How many candles were on the birthday cake? _____

2. On what floor was someone looking out the window? _____

3. What time was it when the accident happened? _____

4. Was the person who was hit wearing stripes or polka dots? _____

5. What was the license number of the parked car on 3rd Street? _____

6. How many windows had curtains? _____

7. How many policemen were at the scene? _____

8. Were there more than five people in the crowd? _____

9. Was the butcher's first name John? _____

10. What was the young person in the crowd eating? _____

11. Was it a two- or four-door car? _____

12. How many trees were in the picture? _____

13. Were there any animals on the street? _____

14. On what street did the accident happen? _____

15. What was the big ad on the wall advertising? _____

16. Was the driver of the car alone? _____

17. What was the number of the shoe repair shop? _____

Memory Makers

Many times we use mnemonic devices or memory aids to help us remember the order of things or a series of things. These mind joggers are named after Mnemosyne, the Greek goddess of memory. See if you can match the mnemonic device with what it is trying to get you to remember. Put the correct letter of the mnemonic answer above the corresponding number at the bottom of the page to spell out a special message.

1. HOMES	A.	notes on the lines on a scale
2. ROY G. BIV	M.	order of mathematical operations
3. DR. MCSNEER	H.	colors of the spectrum
4. FACE	R.	first six Presidents of the U.S.
5. Every Good Boy Deserves Favor	W.	the Great Lakes
6. My Dear Aunt Sally	M.	planets in "order" from the sun
7. King Phillip Crossed Over Fairly Gentle Seas	T.	notes in the spaces on a scale
8. My Very Excellent Mother Just Served Us Nine Pizzas	Y.	continents of the world
9. Andy And Art Pouted Intensely	A.	systems of the human body
10. Wayne And James Made Mud Animals	E.	scientific classification of living organisms
11. 6 A's and an E	O.	oceans of the world

$$\overline{}\ \overline{}\ \overline{}\ \overline{}\quad \overline{}\ \overline{}\ \overline{}\ \overline{}\ \overline{}\ \overline{}\ \overline{}\,!$$

  1   2   3   4   5   6   7   8   9 10 11

Learning Extenders

1. Current research shows that the brain is divided into a right side and a left side (or hemispheres). It is believed that each side is responsible for different functions and tasks. Research right brain/left brain and draw a map of the brain showing what functions are carried out in each side.

2. People try lots of different ways to remember something. Some of those ways include tying a string around a finger, writing it down, associating it with something else they can remember, just to name a few. Interview family and friends and find out how they go about remembering things. Collect data and make a graph showing your results.

3. How we learn and how we remember have a great deal in common. Research the following types of learning and put together a class center explaining their differences.

4. Russian physiologist Ivan Pavlov completed some very interesting experimentation with learning. Research Pavlov and make a shadow box illustrating him performing one of his experiments.

5. There are many memory books on the market today. Locate and read one of them and write a critique of the book. Would you recommend it to someone else? Did it help your memory? What did you like and dislike about it?

6. Construct a "Brain Game" using information about learning and memory that can be played by classmates.

7. Determine the five most important facts you learned about memory. Rank them in order of most important to least important. Explain your rankings.

8. Create a personalized license plate for someone with a good memory . . . for someone with a poor memory.

9. Develop a test that you can give family or friends testing their short-term memory.

10. Write a cartoon style book titled "How to Improve Your Memory." In your creation explain techniques people can use to improve their memory.

11. Make up your own list of mnemonic devices you could use to help you remember important things you are learning in school.

Sweet Dreams

We use the word *dream* to describe many things—dreamboat, dream house, dreamland, just to name a few. What do all of these very different things have in common? They all refer to something ideal, perfect. We think that in our dreams everything will be all right.

To understand dreams, you must first understand something about sleep. You may think that sleep and dreams just happen. Nothing could be further from the truth. Sleep follows a very definite pattern. Each time you go to sleep, your body goes through four different stages. Sleep isn't simply a quiet resting time, interrupted by dreams now and then. Sleep follows this pattern:

Stage One

You begin to drift off to sleep. The brain waves are small and fast, much different than when awake. The heartbeat slows down, temperature drops, and all the muscles start to relax. You change position very often in this stage.

Stage Two

The brain waves become slower, but there are periods of fast waves called "sleep spindles." The eyes may roll from side to side, and you still may be trying to find the most comfortable position. If someone woke you during this stage, you would probably say you were not asleep. This lasts about thirty minutes or so.

Stage Three

About thirty to forty minutes after falling "asleep," you enter stage three. This is a deep sleep. The brain waves are bigger and more spread out. Your blood pressure drops and everything is completely relaxed. It takes a loud noise to wake you.

Stage Four

The deepest sleep occurs here. The brain waves are very slow, large, and regular. Your body goes through very few movements, and even loud noises may not wake you. During this stage people will sometimes "talk in their sleep" or sleepwalk. Usually, most of the first half of the night is spent in stage four sleep, especially if you are tired.

After this stage, the process reverses and you gradually go from stage four to stage three and then stage two. When you return to stage one, your eyes start making rapid, darting movements, as if you were looking at something. These movements, called REMs (rapid eye movements), are the beginning of dreams. About ninety minutes have gone by. This dreaming stage will take about twenty minutes before you drift back into stage two and continue the up and down pattern again. The entire pattern is repeated four to six times each night. If you were awakened at this point, you would probably remember your dream. But if you are not awakened, you might not even remember having had a dream.

As the night goes on, the dreaming periods become longer and the deep sleeping stages are shorter. Toward morning, dreams may last more than an hour.

A newborn infant spends about half of its sleeping time in REM and the adult about one-sixth. This means that someone who sleeps seven to eight hours spends about one and one-half hours dreaming. If sleeping pills or alcohol is taken, a person may not dream. However, when this is stopped, the person dreams almost continuously, as though he or she needs to catch up on dreaming. Dogs, cats, and all other mammals dream every time they fall asleep, and over eighty dream in color!

In ancient times, it was believed that dreams foretold the future. The Egyptians thought dreaming was so important that they built temples to Serapis, the god of dreams, with the hope of stirring helpful dreams. They even had "dream books" which were used to interpret the meaning of dreams.

The Iroquois Indians had a dream-sharing gathering every winter. In ancient Greece, sick people traveled to holy places to sleep, in hopes of having dreams that would tell them how to get well.

In modern times, the best-known theory of dreaming came from Sigmund Freud in his book *The Interpretation of Dreams*. Freud was a physician in Austria and was well trained in the workings of the nervous sytem. From his experiences, he came to the conclusion that many ailments, both mental and physical, have their roots in childhood experiences. He used dreams in his treatment of patients believing that these hidden thoughts would come out in the dreams, and by analyzing them many problems could be solved.

Recently, some scientists have suggested that dreams are simply a way of processing information. During the day, we are all bombarded with too many facts and experiences to deal with, and dreams are just a way of sorting the important from the unimportant. Take Jack Nicklaus, for example. He told a reporter that he couldn't understand why his golfing scores were going down until he dreamed he was holding his club in a slightly different way from his normal grip. The next day he swung the club as he had dreamed and his scores improved. Robert Louis Stevenson reported that it was from a dream that he got the idea for *Dr. Jekyll and Mr. Hyde*. Whatever the case and whatever you think about dreams, the next time you fall asleep, sweet dreams!

In Your Dreams

We spend a lot of our lives asleep. Come to think of it, we spend a lot of our time dreaming, eating, watching television, and going to school! Using the "information" below, compute (with a calculator if you choose) the answers to the problems. Find the correct answer in the answer bank and write it above the corresponding number at the bottom of the page. When you have answered all the questions, complete the final problems at the bottom of the page. Your final answers should be 170 and 2272.5.

(*All numbers are averages and estimations. They may or may not be accurate for every individual.)

Sleeping*

one- to five-year-old = 14 hours
five- to ten-year-old = 12 hours
ten- to fourteen-year-old = 10 hours
fourteen- to eighteen-year-old =
9 hours
eighteen to adult = 8 hours

Dreaming*

Rapid eye movement (REM), during which dreaming takes place, lasts about 20 minutes and occurs from four to five times a night for a total of about 90 minutes per sleep period.

Eating*

Breakfast = 15 minutes
Lunch = 25 minutes
Dinner = 35 minutes

School*

Kindergarten = 540 hours per year
Grades 1 through 12 = 1080 hours per year (6 hours per day)

Watching Television*

Weekdays = 3 hours per day
Weekends = 6 hours per day
(1 year = 52 sets of weekdays and weekends)

1. How many hours would a one-year-old sleep during the month of July?

2. How many minutes do you dream each week?

3. In a normal year, how many more hours would you watch TV than go to school?

4. How many hours would a twelve-year-old dream during the month of June?

5. How many minutes does it take the average person to eat each week?

Answer Bank
(Don't worry about the label for the answer. We're just concerned about the number this time!)

140
434
45
3.75
2570
324
525
42
630

$$[(\underline{}\ \underline{}\ \underline{} + \underline{}\ \underline{}\ \underline{}) - \underline{}\ \underline{}\ \underline{}] - \underline{}\ \underline{} - \underline{}\ \underline{}\ \underline{} = 170$$
$$1 2 3 4 5$$

1. How many more hours a year (365 days) would a ten-year-old sleep than go to school?

2. How many more minutes a week does it take you to eat dinner than breakfast?

3. How many more hours a week does a one-year-old sleep than a twenty-one-year-old?

4. How much time is left on a Monday for a twelve-year-old after eating, sleeping, watching TV, and going to school?

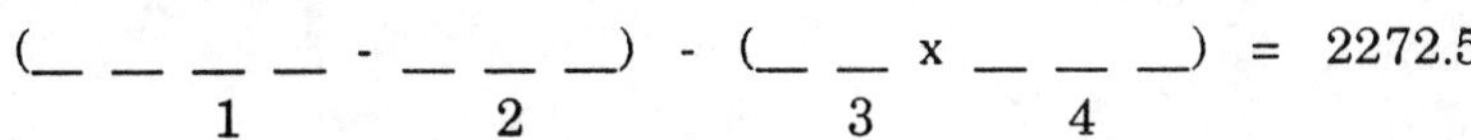

$$(\underline{}\ \underline{}\ \underline{}\ \underline{} - \underline{}\ \underline{}) - (\underline{}\ \underline{} \times \underline{}\ \underline{}) = 2272.5$$
$$1 2 3 4$$

Author, Author!

One of the most important research skills to learn is how to use the card catalog in the library. Here's an activity to check your skill. Pretend you want to find a book about dreams. When you look up that subject you might find the following four cards. Use the information on these cards to answer the questions given below to complete the old saying, "A dream not understood is like a letter _ _ _ _ _ _ _ _!

<table>
<tr><td>

DREAMS

j.   Stafford, Patricia.
154.63   Dreaming and dreams/Patricia
STA   Stafford; illustrated with photographs and diagrams by the author–1st ed. New York, Atheneum; 1992
     53 pp.
     Includes bibliographical references and index.
     Discusses dreams, remembering them, controlling them, why we dream, and dream themes and meanings.

</td><td>

DREAMS

j.   Parker, Steve
154.63   Dreaming in the night: how you
PAR   rest, sleep, and dream/Steve Parker, New York, F. Watts, c 1991
     32 pp.
     Includes bibliographical references and index.
     Examines dreaming as a neurological function and describes the change in the nervous system.

</td></tr>
<tr><td>

DREAMS

154.63   Hirsch, S. Carl.
HIR   Theater of the night: what we do and do not know about dreams/S. Carl Hirsh.–Chicago, Rand, McNally, c. 1976
     123 pp: illus.
     Examines past and present attitudes toward the phenomenon of dreaming and current dream research.

</td><td>

DREAMS

154.63   Lindsay, Rae.
LIN   Sleep and dreams/Rae Lindsay; illustrated by Leigh Grant.–New York, Watts, 1978
     63 pp.; illus.
     Discusses the stages of sleep and sleep disorders.

</td></tr>
</table>

1. Whose book was not published in New York?
   U. Hirsch   M. Lindsay   S. Stafford
2. Which book was printed in 1991?
   L. Sleep and Dreams   O. Dreaming and Dreams   N. Dreaming in the Night
3. Which book is the longest?
   E. Sleep and Dreams   O. Theater of the Night   G. Dreaming and Dreams
4. In the card catalog, which book would come first (when listed alphabetically by author)?
   E. Dreaming and Dreams   P. Theater in the Night   L. Sleep and Dreams
5. If I couldn't get to sleep at night, whose book might I read?
   E. Lindsay   P. Stafford   O. Parker
6. How many of these books are illustrated?
   S. two   I. one   N. three
7. Whose book has photographs?
   E. Stafford   L. Lindsay   M. Parker
8. Which book would help you learn to keep a "dream journal"?
   S. Dreaming in the Night   R. Sleep and Dreams   D. Dreaming and Dreams

Learning Extenders

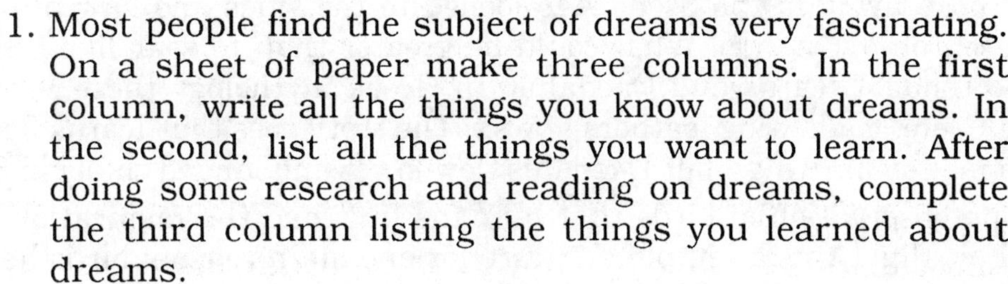

1. Most people find the subject of dreams very fascinating. On a sheet of paper make three columns. In the first column, write all the things you know about dreams. In the second, list all the things you want to learn. After doing some research and reading on dreams, complete the third column listing the things you learned about dreams.

2. Keep track of your dreams for two weeks in a dream journal. See if you can figure out why you might have dreamed what you did and what these "symbols" have to do with the "real world."

3. Compose a poem or song about one of your favorite dreams.

4. Research the subjects of sleepwalking and talking in your sleep. Why do these two activities happen? What does it mean? Are these activities dangerous? Is there any way you can control them? Report your findings to your class in a poster called "In Your Dreams."

5. Invent a machine that could help you remember your dreams. If it is something you could construct, make one to share with your classmates.

6. Dreams have played an important part in our history. Read about five famous dreams and rank them as to which you think had the most important effect on history. Give reasons for your choices.

7. Develop eight to ten questions on dreams and survey the members of your class. You might want to ask questions like "Do you dream every night?" "What are some of the things you dream about?" "How do you remember your dreams?" Construct a graph or chart to share your findings.

8. Sometimes night is a scary time for children. To help them feel better about night and dreams, write a book for primary students. What are some of the things you would want them to know about dreams? See if you can share the book with younger students in your school.

9. On occasion people have nightmares. Research the subject of nightmares and make a semantic map (web) of your findings.

nightmares

It'll Never Fly

Ever since those living in the Stone Age looked to the skies and saw birds at wing, flight has been on our minds. What could be greater than to soar like a bird in the heavens? So thought Icarus and Daedalus. In Greek mythology these two, trying to escape from Crete, made wing feathers of wax. The story goes that Icarus flew too high and the sun melted his wings, but Daedalus flew lower and landed safely in Sicily.

It is from the observation of birds that humans first had the inspiration to fly, but imitating their flight was a complete failure for one main reason: birds have a lot of muscle power for their size, lifting very little weight. Our muscles are very puny compared to theirs and our bones are solid while theirs are hollow. It's not as simple as strapping on a pair of wings and flapping until we get airborne!

Throughout the Middle Ages, scientists and inventors dreamed up ideas for "flying machines." In 1488 Leonardo da Vinci made a sketch of a flapping-wing machine, or ornithopter. It was never built, but had it been, its six hundred pounds would surely have kept it on the ground.

There are many other historical flying stories. In 1678, the Frenchman Besnier was supposed to have made a gliding flight using a pair of large, wooden wings. (It was not unlike the paper airplane gliders you may make to "fly" your trash from your desk to the trash can!) In 1709 Portuguese priest Laurenco de Gusmao was said to have designed the "Great Bird" whose lifting power was believed to be powerful magnets. There was no limit to the creativity. Hot-air balloons and gliders had been flown since before the 1900s, but they were unreliable and could not carry a person over a long distance. It wasn't until the Wright Brothers in 1903 that the first controllable air flight was made and what a flight! It went 120 feet (37 meters) and lasted 12, count them, 12 seconds!

Much has changed since that December day at Kitty Hawk. Today, a "jumbo jet" fully loaded with fuel and some 500 passengers weighs more than 350 tons, and yet it can take off and fly. Isn't it amazing how that can be done?

The key lies in four terms–lift, drag, thrust, and weight (gravity). The airplane must have enough **lift** and **thrust** to overcome **weight** and **drag**. But what does all that mean for those of us about to take off on an air flight?

Because an airplane is a craft heavier than air, it can fly only if enough lift is produced that will overcome the forces of weight. Thrust is required to move the airplane through the air and is produced by the plane's engine. It must be stronger than drag which is the resistance of the air to anything moving through it.

Does all this sound complicated? Well, it really isn't. You experience these same four forces every time you ride a bicycle. You pedal, producing thrust by turning the rear wheels against the ground. The air pushing against you is the drag. If you place one hand out with the palm down and slightly tilted to face the oncoming air, your hand will rise, producing lift. Gravity (or the weight of your bike) is what is keeping you earthbound.

For an airplane, the key to getting in the air and staying there lies in the cross-sectional shape of the wings–rounded at the front, flat underneath, curved on top, and tapered sharply to a point at the rear. This shape is known as an airfoil. The wings push the air aside as they move forward, and, as the air passes over the curved top surface, it speeds up. When air moves faster, its pressure drops, and the wings tend to lift.

You can see how lift is created by doing a little experiment. Cut a piece of paper two inches wide and seven inches long (5.08 x 17.78 cm). Hold it against your chin under your bottom lip with the narrow part. Blow hard over the top of the paper and watch it rise. What actually happens is the "air in a hurry" on top of the paper has less pressure, and the pressure under the paper is greater which lifts the paper up. This is known as Bernoulli's Law.

Thrust, the forward force that keeps the plane moving through the air, is produced by the engines. Most jet engines produce thrust by burning kerosene fuel and ejecting the hot gases to produce high speed. As the gases shoot backward, the plane moves forward.

Drag, or air resistance, must be kept at a minimum to allow for an economical flight. This is done in the design of the aircraft. If you look closely at a plane, you notice that there are as many smooth and slippery surfaces as possible. There are no raised or rough surfaces to block airflow. Automobiles also try to reduce the amount of drag in their designs, making driving them more fuel efficient.

The pilot has many "aids" at his or her disposal. On the wings there are ailerons which are used to make the plane dip or rise. The rudder can be found on the tail of the plane and is used to turn the nose of the plane right or left. To make the plane go up or down, the pilot operates the elevator, also located on the tail. In addition to these, there are many instruments and controls which must be watched carefully if the flight is going to be successful.

So the next time you see a plane in the sky or board one for a vacation, think about all the scientific processes that are going on in order to make that flight happen!

GA1512

Take Flight!

Thanks to air travel, you can get from one city to another in a very short period of time. In fact, an average 747 travels at 500 miles per hour (give or take a mile or two). The map below includes the various time zones for the United States as well as mileage between several cities. Using this information, answer the ten questions below by selecting the correct answer from the answer bank. Write the letter next to the correct answer above the corresponding number at the bottom of the page to spell out the answer to the question, "What do you call an almond that travels in space?"

Air Distances and Standard Time Zones

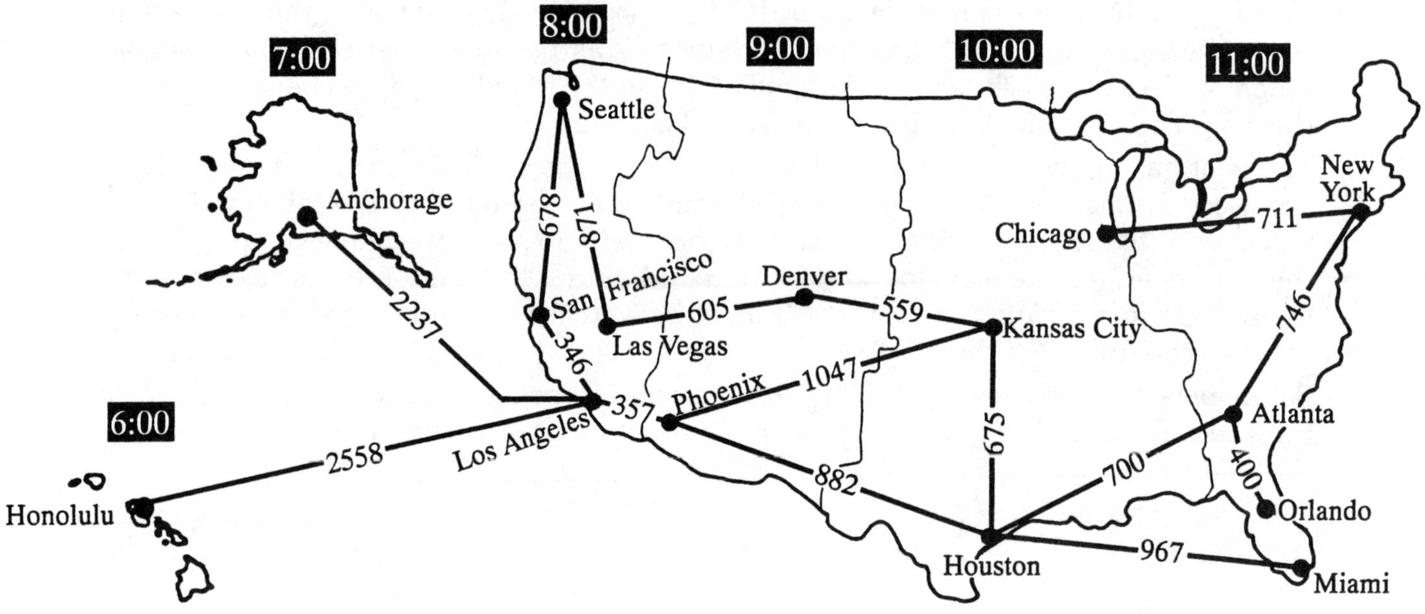

1. If you leave New York at 11:00 a.m., about what time would you arrive in Houston?
2. What's the air distance from Seattle to Kansas City through Las Vegas and Denver?
3. Taking the most direct route on this map, how far is it from Miami to Los Angeles?
4. How many minutes less would it take to fly from Anchorage to Los Angeles than from Honolulu to Los Angeles?
5. What's the round trip mileage from Seattle to Phoenix if you travel through San Francisco and Los Angeles?
6. You leave Kansas City at 10:00 a.m. and must personally deliver a package to the airport in Houston and then one in Phoenix. If there are no layovers, about what time would you get back to Kansas City?
7. According to this map, how many air miles is it from Epcot Center to Chicago?
8. If it's 9:00 p.m. in Atlanta, what time is it in Honolulu?
9. If you leave Phoenix at noon, what time would you arrive in Kansas City?
10. How long would it take to fly from San Francisco to Los Angeles?

Answer Bank

A = 12:55 p.m.

A = 2206 miles

S = About 38 minutes

N = 4:00 p.m.

O = 1857 miles

T = 2762 miles

T = About 42 minutes

U = 1:05 p.m.

R = 3:13 p.m.

N = 2035 miles

<u>   </u> <u>   </u> <u>   </u> <u>   </u> <u>   </u> <u>   </u> <u>   </u> <u>   </u> <u>   </u> <u>   </u>
  1   2   3   4   5   6   7   8   9   10

It's as "Plane" as the Nose on Your Face

Homophones are words that sound alike but have different meanings and may be spelled differently. See if you can "land" the pair of homophone answers to these clues. Take off.

1. A man went to borrow money from a bank by himself. He was trying to get a ___________ ___________.

2. A thief took the metal from the mill. He was trying to ___________ ___________.

3. She worked hard to get money to buy a vase. She was trying to ___________ an ___________.

4. She went to the top of the mountain to take a quick look at the view. She wanted a ___________ ___________.

5. She mailed her mother a perfumed envelope. It was a ___________ ___________.

6. The teacher did not permit anyone to talk. In her class speaking ___________ was not ___________.

7. He paddled the boat down the highway. He ___________ on the ___________.

8. They are charging us too much to take the bus. It's not a ___________ ___________.

9. He cut his hand on a broken window glass. It was a ___________ ___________.

10. The cold wind made his face turn a different color. It ___________ till he was ___________.

11. The entire class fell into the pit. It was a ___________ ___________.

12. The carpenter's material put everyone to sleep. It was a ___________ ___________.

13. The group of cattle was making a loud noise that we could pick up. We ___________ the ___________.

14. Dad's piece of fishing equipment wasn't fake. It was a ___________ ___________.

15. She tried to grab the bodies of water. She wanted to ___________ the ___________.

Learning Extenders

1. It takes a lot of training to be an airplane pilot. Contact (either by phone or letter) a pilot in your area and interview him or her. Make sure that you think about all the questions you want to ask before the interview. Present your interview to your class in the form of a radio broadcast using a tape recorder. (If you have to do your interview by letter, have another adult answer the questions you asked the pilot.)

2. Airplanes are very complicated pieces of "machinery." Research some of the equipment that is used to fly the plane, and put that information on a poster titled "Up in the Air!" that can be shared with the rest of your class.

3. Construct a time line listing the important historical events that have taken place in aviation history.

4. Investigate how a hot-air balloon works and try to make one out of recycled materials.

5. The "modern" passenger airplane has changed a great deal since the Kitty Hawk. Show those changes by drawing planes from that time until now.

6. Write a poem or song titled "If I Could Fly."

7. Have a class paper airplane contest. See who can "design" a plane that will fly the highest, that will go the farthest, and that will land the smoothest. Put the winners on display for the rest of the school to see.

8. Collect and display plant seeds that "fly." Explain about each seed and the scientific principles that are involved.

9. Write to your local NASA center and, from the materials you receive, put together a class center titled "Traveling in the Future."

10. Write an imaginary newspaper account of a famous first flight. Make sure that you use the who, what, when, where, and why of journalism.

Holograms

Most everyone has used a camera and taken a picture. When these photographs are developed, you are able to see a two-dimensional picture of your subject. But wouldn't it be great if you could take a picture where you can see not only the front but also the sides and back of your subject? Sounds like science fiction? Well, it's not. It's holography. The word *holography* comes from two Greek

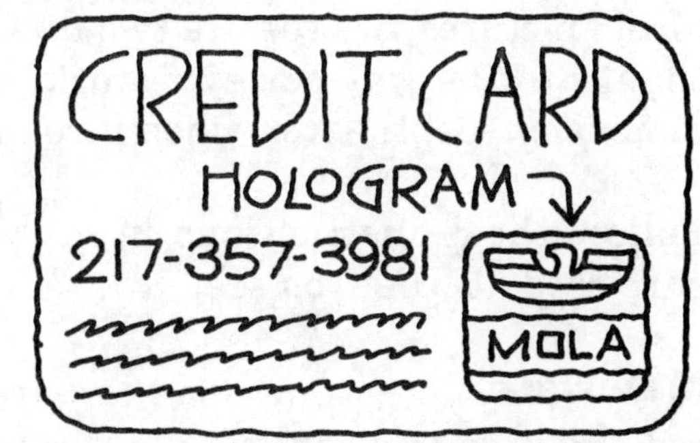

words that mean "entire picture" and was discovered in 1947 by David Gabor, a British electrical engineer. He discovered that three-dimensional images could be recorded on photographic plates. This means that when viewed, the image changes when you look at it from different angles. But it wasn't until 1961 with the invention of the laser that modern holography was developed. The laser was important because it produces "coherent" light. What does all this mean?

Holography uses an effect in physics called interference. In interference, two coherent waves with the same wavelength combine together. These waves can combine in different ways. You probably already know that light travels in waves. The top of the wave is called the crest and the bottom the trough. In a hologram, the crests and troughs of each wave can come together and form a combined wave as bright as the two are separately. Or the crests of one can combine with the troughs of another, and the two will cancel each other out producing a pattern of light and dark patches. This is called an interference pattern.

In holography a beam of light from a laser is split by a semitransparent mirror. One beam, called the object beam, lights up the subject to be recorded. This part of the beam then hits a photographic plate–a glass plate with light sensitive emulsion on one side. The other beam, called the reference beam, is reflected directly onto the plate. The two beams coincide to create an interference pattern, which is recorded on the photographic glass plate.

To view this scene, if coherent light from a laser is shone onto the plate from behind and you look through it, you see the original object in three dimensions. If you move your head sideways while looking at the object, the picture changes just as it would if you were looking at a solid object. (White light cannot be used in this process since it is a mixture of different wavelengths. In other words, white light is not coherent.)

Holograms can now be found on credit cards to discourage counterfeiting, on magazine covers to present three-dimensional images, at checkout counters to scan the bar code on products, and in aircraft to allow pilots to read instruments without taking their eyes off the target. What is the future of holography? Research is now being done to create a three-dimensional color television. Since lasers can be focused to give very small "pinpoints" of light, holograms can be used to store large amounts of information and someday may replace microfilm as a way of storing vast amounts of information.

A Third Dimension

Most "pictures" we see are two dimensional–they have height and width. However, thanks to the way our eyes work, sometimes we can fool them into seeing a third dimension–depth–even when it really isn't there.

Follow these instructions in making a set of the 3-D glasses and you may be surprised at what you see!

Materials:
red and green cellophane
cardboard or cover stock paper
scissors
glue
red and green crayons or colored pencils

Procedure:
1. Using the pattern below, draw a pair of glasses' frames on cardboard or cover stock.
2. Cut out the glasses and place a piece of red cellophane over one eyehole (A) and a piece of green over the other (B).
3. Fold the glasses on the double line and glue the two sides together. (Make sure you don't get any glue on the cellophane!)
4. Draw a picture using a red pencil and another picture with a green pencil. You might even want to make a third picture using both colors.
5. Look at each picture. Record what you see when you close one eye, then the other. Do you see a 3-D effect when you use both eyes and look at the red and green picture? Why do you think these things occur?

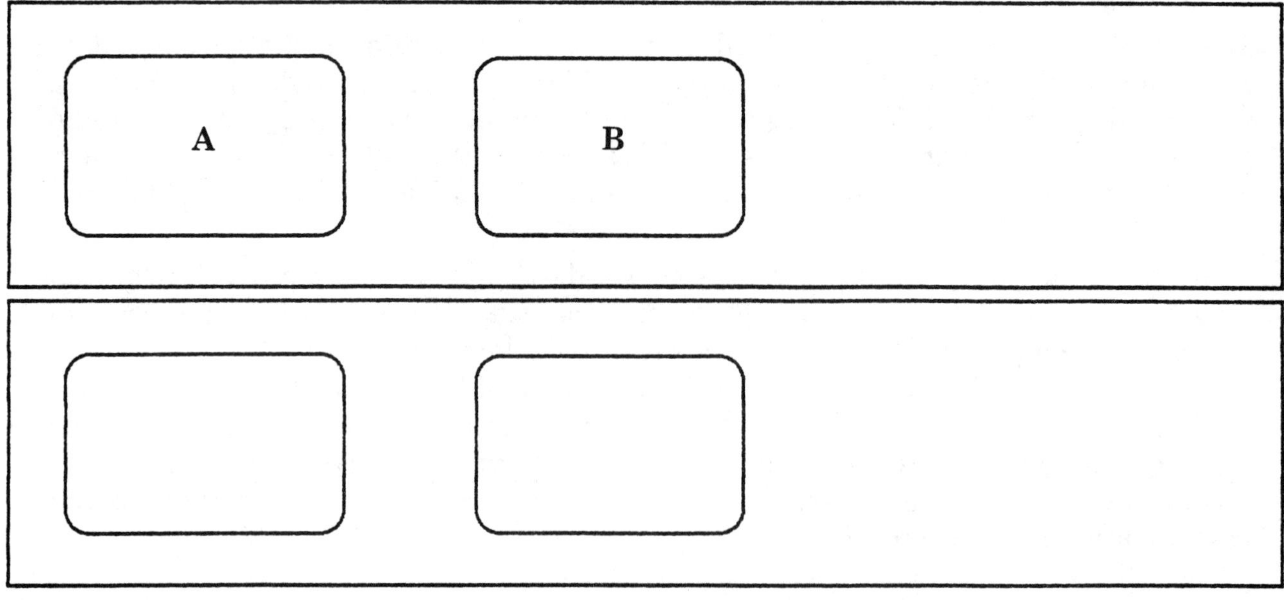

A Compound Problem!

Much of our English vocabulary is derived from Latin and Greek. Words that are made up of one or more words are called compound words. In Column 1 you will find a list of some "scientific" compound words. Match each with its correct "definition" in Column 2. By each definition is a letter. Write that letter at the bottom of the page above the number that corresponds to the scientific term. You will spell out the answer to the question "What do you get when you look at a hologram?"

<table>
<tr><td>

Column 1

1. Hologram
2. Biology
3. Photography
4. Hydrophobia
5. Megalith
6. Microscope
7. Telephone
8. Perimeter
9. Asteroid
10. Transplant
11. Chromosomes
12. Dinosaurs
13. Biodegradable
14. Seismology
15. Atmosphere
16. Millimeter
17. Thermometer
18. Forecasting

</td><td>

</td><td>

Column 2

E = heat, measure
A = star, shape
N = life, word
I = monstrous, lizard
V = vapor, sphere
N = life, decompose
E = large, stone
G = earthquake, word
I = thousand, measure
N = across, set firmly in position
A = whole, small weight
H = through, measure
V = water, fear
W = before, to form
G = color, body
R = small, target
E = light, to write
C = distance, sound

</td></tr>
</table>

__ __ __ __ __ __ __ __ __ __ __ __ __ __   __ __ __ __
1 2   3   4   5   6   7   8   9 10 11 12 13 14   15 16 17 18

GA1512

Learning Extenders

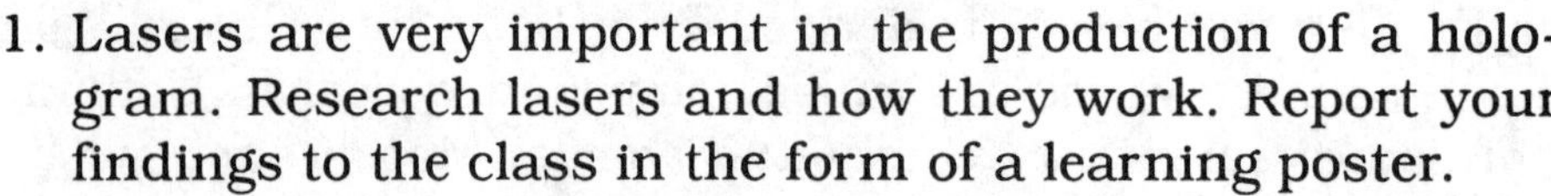

1. Lasers are very important in the production of a hologram. Research lasers and how they work. Report your findings to the class in the form of a learning poster.

2. Design a series of diagrams to show how a hologram is made.

3. Lasers are used in many different occupations. Using the yellow pages for your community, see if you can find three places that use lasers. Contact these places to see if you are correct.

4. Locate as many examples of the use of holograms as you can. You can find them on magazines (like *National Geographic*), on trading cards, on credit cards, just to name a few locations. Put together a classroom display titled "Would You Look at That!"

5. If your teacher or a parent has access to a laser, try to make a hologram. It's not all that difficult. Maybe you could contact a high school teacher or student to help you.

6. Select a book on holography like *The Hologram Book* by Joseph Kasper and Steven A. Feller (Prentice Hall, 1985) and complete the activities below.

   a. Summarize the main ideas or facts found in the book.

   b. Select several key words or terms from the book and classify them in some way.

   c. Compare your book with another book on lasers. How are they alike and how are they different?

   d. Suppose you were to write a new book on lasers. Create an original book jacket for your book.

   e. Would you recommend this book to someone else? Why or why not?

7. Write a business letter to the Holography Information Center, P.O. Box 586, Lake Forest, Illinois 60045 (or to another organization that can provide information on holography) and put together a learning center for your classmates.

8. You have just been hired as the Public Relations Manager for the Laser/Hologram Company. It seems that the hologram has been entered in the "Most Important Invention of the Century" contest. It is your job to create a P.R. packet to help win that award!

Mysteries of Insect "Communities"

Back when humans first inhabited the earth, they were pretty independent. Most of the day was spent hunting for food and keeping from being some animal's dinner! As time passed, we became more social creatures. We learned that there were advantages to living in groups. We also discovered that life was easier if work could be shared and then no one person had to do everything. From this, we developed communities.

Many, many years before we grouped ourselves together, insects had already discovered the advantages of organized living. Ants and honeybees are just two of the many insects that have a very structured life.

Ants

If you have ever gone on a picnic, you've seen ants. First there is one poking his nose around in your food and, not long after that, there are thousands carrying off whatever they can.

Ants belong to a special family of insects, the Formicidae. If you look closely at an ant, you see a tiny extra segment between the throat and the stomach (thorax and abdomen). Ants also have large jaws which they use for almost everything, like cutting up seeds, capturing other insects, digging in the ground, and making tunnels in wood. There are about 8000 different kinds of ants, and they live in almost all parts of the world except for the coldest regions.

These creatures live together in colonies that sometimes contain thousands of individuals. In a typical colony there are usually three kinds of ants. These are the queen, many worker ants, and a few males. Sometimes there are also soldier ants.

The female plays the most important role in an ant community. All the worker and soldier ants are female, but they do not lay eggs. Males are found only during certain times of the year. It is at this time that new queens are produced and when "swarming" happens. During "swarming" the new queens and the males leave the nest to mate and start new colonies. Once the mating is complete, the males die. When the new queen reaches her new community, she sheds her wings.

Only the males and queens have wings. All the soldiers and workers are wingless. Worker ants are the ones you might see crawling about when a nest has been disturbed. They do all the work of nest building, gathering food, and caring for the young. They have very short life spans—a few weeks or months. The queen, however, may live for five or six years.

Of the 8000 species of ants, many have ways of making a living—not unlike humans. There are weaver ants, hunter ants, and farmer ants. Yes, farmer ants. These ants actually grow crops as a source of food. Other harvester ants just gather their food, seeds.

Most anthills in the Western plains of the United States contain harvester ants. Inside the hill there are many chambers and tunnels. Some of the tunnels are filled with seeds and others contain young ants. All the chambers are connected by passages. Those passages near the top of the hill are used for raising the young ants. Deeper in the ground are chambers where ants spend the winter. Deeper yet are chambers for food storage. So, not only are the ants organized in communities, their "homes" are also organized for the greatest efficiency.

Honeybees

Like ants, bees have a very structured existence and as many as 50,000 may live in one community. In each "society" there are usually three castes or types: queens, workers, and drones. As a larva (the wingless, worm-like form of a newly hatched insect), the queen is fed a special food called royal jelly. As mother and head of the colony, she develops faster and grows larger than her "subjects." After mating, the queen will lay about 3000 eggs each day, each in its own individual cell. Some will hatch into queens, others workers, and the unfertilized eggs will become drones.

Worker bees are also females with imperfectly formed reproductive organs. They have specialized mouths used to gather nectar and mold wax. Their hind legs are equipped for carrying pollen from plant to plant–an activity that is very important in nature. Another job of the worker bees is to protect the queen and defend the colony even if it means giving up their lives.

When a worker bee discovers a supply of nectar, it flies back to the hive to tell the other bees exactly where it is. It also tells them what kind it is and how good it is. How can this be, you ask? The female worker bee returns to the hive and dances. Yes, dances. The pattern is a figure eight and is repeated over and over as the other female bees watch. Through research, scientists have discovered that the most important part of the dance is the straight run through the middle of the figure eight. This run shows the direction from the hive.

If the bee is dancing outside the hive, she lines up with the sun and turns to point toward the food. If she is in the hive, she will point her head up (as if the sun were overhead) and turn right or left to show where the food is. As she runs through the figure eight, she will wag her head and tail from side to side. The further away the food is, the faster the wagging.

As the worker dances, her wings vibrate so fast that they buzz. The other worker bees touch the dancer with the antennae to feel the vibrations. They also sample a drop of the nectar she has found. Within a few minutes the first bees figure out where the food is and they are off. The information given is so accurate that scientists have been able to follow the directions and find the same flowers!

Drones are the male honeybees whose only purpose is to fertilize the few young queens each fall.

When the population of one bee "community" becomes too large, swarming will occur and a new city will be started. The first swarm includes a queen and many workers, leaving the old colony to be ruled by a new queen.

46

Abstract Insects

Just think. If you crossed a girl with a frog, what might it look like? Or an ostrich with a horse? They would look pretty funny, huh?

Well, what about crossing one insect with another? Say a spider with a wasp? Or a grasshopper with a firefly? The combinations of possibilities are endless.

Select two unlikely insects and "bring them together" to form a new insect. When you are finished with your drawing be sure to do the following:

1. Name your new insect.
2. Label its parts.
3. Tell where it lives and what it eats.
4. Write a short story about a day in its life.

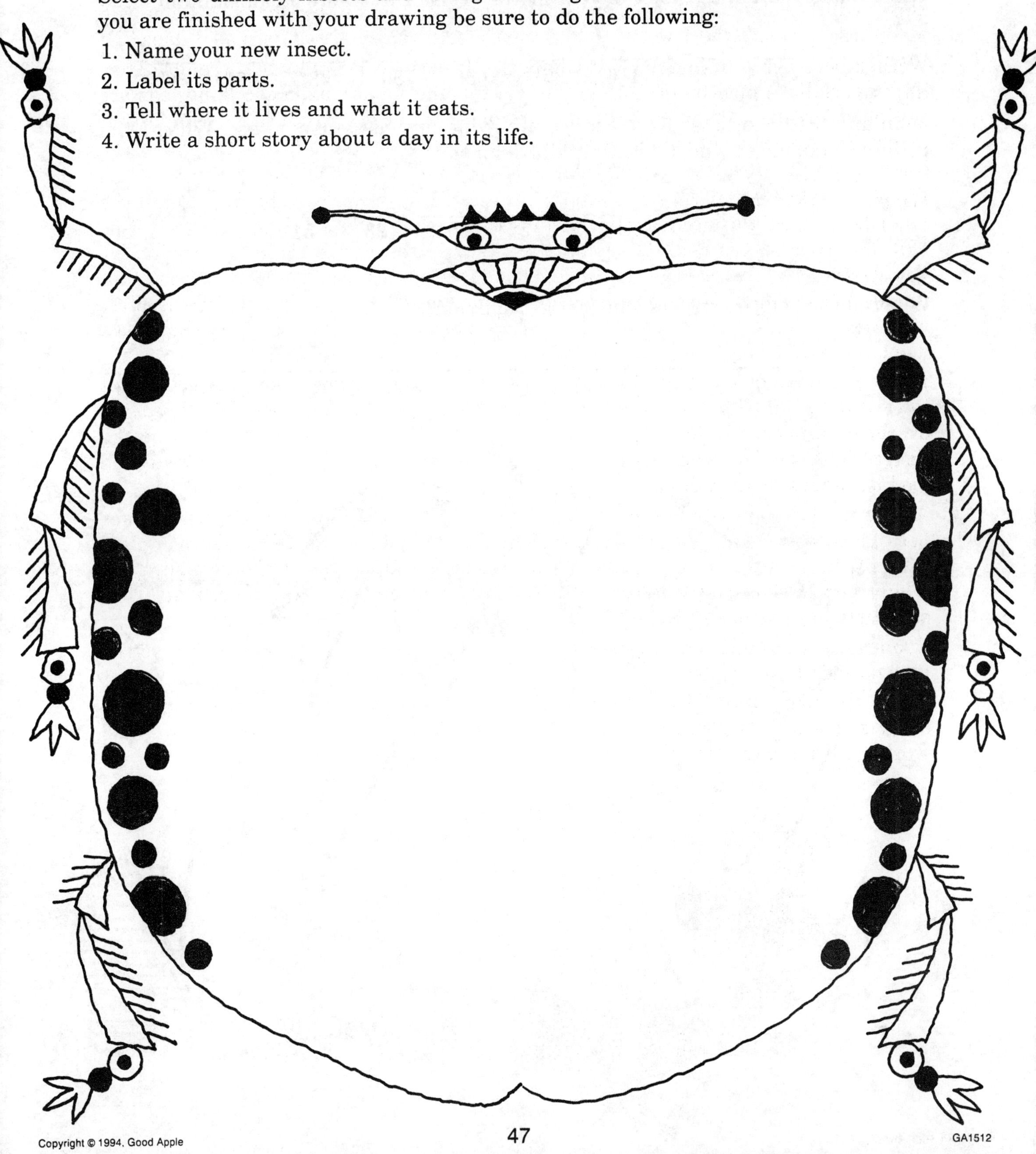

GA1512

Don't "Bug" Me with Your Questions!

Measure the angles below and then find each measurement in the answers to the three questions. Each time the answer appears, write the vertex letter of that angle above it to spell out the correct responses to the "bugging" questions.

What would you call the home of your mother's sister if it were on a mountain?

————— ————— ————— ————— ————— ————— ————— ————— ————— —————
126°   37°   126°   116°   37°   108°   80°   23°   86°   86°

John and his little sister Kathleen were playing and Sis broke a vase. When Mom got home and asked who did it, what did John say?

"————— ————— ————— ————— ————— ————— —————"
  61°   126°   108°   55°   51°   23°   51°

What do you call two spiders who just got married?

————— ————— ————— ————— ————— ————— ————— ————— ————— ————— ————— —————
108°   80°   30°   37°   30°   150°   86°   55°   150°   30°   135°   47°

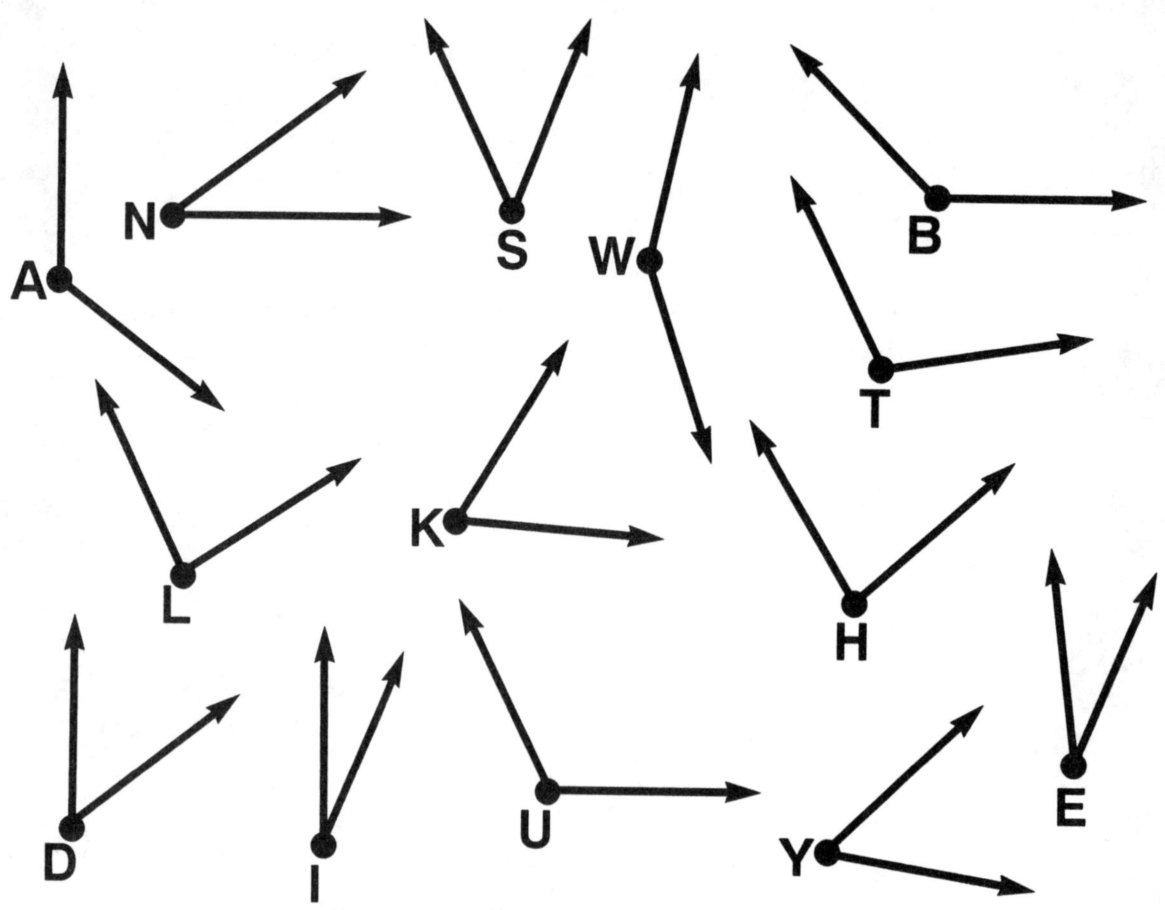

Learning Extenders

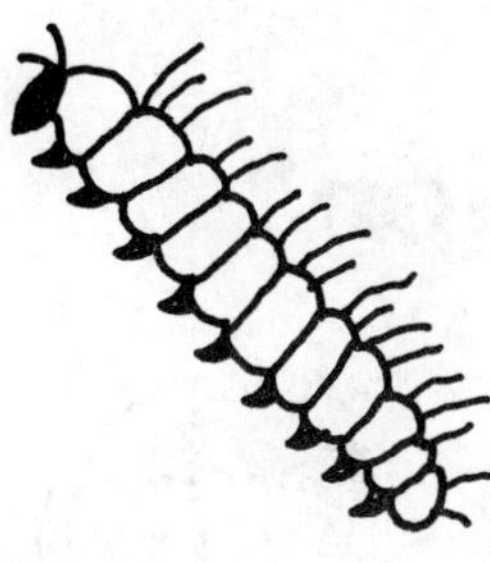

1. Create a diorama (shadow box) of a scene showing how bees live together in a community.

2. Construct an ant farm that you and your classmates can use to observe ant behavior. Keep an account of your observations in your "ant journal."

3. Write a children's picture book titled "Insects That Live in Communities."

4. You have just been hired to illustrate the covers of the following books:
   Bees Do It: Living and Working Together
   My Aunt's Ants
   What would your covers look like?

5. There really are killer ants and killer bees! Research these two insects and list ten facts about each. On a world map, show where these two insects can be located.

6. Prepare a television talk show for your classmates. Interview "experts" (other classmates who have done research) on the subject of insect communities.

7. Make a collage or mobile of insects that live in communities.

8. Select a book on insect communities and do the following:

   a. Summarize the main ideas or facts found in the book.

   b. Select several key words or terms from the book and classify them in some way.

   c. Compare the book with another book on the same subject. How are the books alike and different?

   d. Would you recommend this book to anyone? Give five reasons for your choice.

9. After researching several insects that live in communities, determine which one you think has the most formal community living arrangement. Be sure to justify your choice with facts from your reading.

10. Select one insect that lives in a community. Find out about each member of that community (for example, bees: queen bee, worker bees, etc.). Design diagrams showing the different body structure of each member and label each part.

Mysteries of Animal "Communication"

If we want to communicate with someone, what do we do? Talk, write a letter, use "body language"? There are many things we can do to let someone know how we feel or what we are thinking. But what about animals? How do we communicate with them?

If you've ever gone to the circus or a sea show, you've seen dogs do exactly what they are told, parrots answer questions, dolphins respond to specific commands, and apes act almost human. How do they do that? Are people really "talking to the animals"?

Although all animals have some way of communicating, humans are (as far as we know) the only animal that actually uses language. Or are we? Many scientists argue that we can't be completely sure that animals don't use language and even if they don't, that doesn't mean that they can't learn how to.

All language has two things in common: a group of symbols each with specific meaning and a set of rules to tell how these symbols should be put together. Part of using language is being able to understand the meaning of words (or symbols like in sign language). But a person must also be able to understand the rules by which words are put together. For example, the sentences "Mary gave Jane a gift" and "Jane gave Mary a gift" are made up of exactly the same words but mean completely different things. Because you know the "rules" for word order, you can figure out who gave the gift and who received it. In order for an animal to understand a language, it would also have to learn the rules as well as the symbols.

So you may think that means because your dog responds to you, that he or she understands your language. More likely it's not the words you are using, rather the tone of your voice and your body movements to which the dog is responding.

It does seem, however, that there are two animals–dolphins and apes–that may be able to understand both symbols and rules. Both have "learned" a language (symbols and rules) in communicating with people.

Apes

In June 1966, two teachers began raising the chimpanzee Washoe as if it were a human child. Their goal was to see if they could bridge the language gap between humans and apes. After studying tapes of chimpanzees "talking," they realized that they could lip-read what the chimp was saying even though the sound being made didn't resemble English. They decided that maybe it was the vocal apparatus that made it difficult for apes to speak, not that they couldn't learn a language.

Washoe's "parents" used a modified version of American Sign Language (ASL) to communicate words and ideas rather than talking to her. In ASL, word order does not make any difference like it does in English, and much of the meaning is sent through facial expressions.

Washoe learned how to get herself dressed and undressed, how to brush her teeth, and how to sit at a table and eat with a fork and spoon. After about four years, Washoe could use about 132 signs. She knew a dog was a dog regardless of where she saw it—it didn't make any difference what breed or color it was—she still recognized it. The work with Washoe, however, did not meet the strict scientific standards for "experimentation" and was strongly criticized.

In 1972 a Stanford University graduate student began to teach ASL to one of the most famous gorillas—Koko. According to Koko's "parents," she learned over 500 signs and could make statements up to six signs long. Not only could Koko talk about objects, she would also describe her feelings, make jokes, rhyme words, and insult people! She even had a cat for a pet! Some scientists questioned if Koko really knew what she was doing, so one asked her if she were a person or a gorilla. Koko responded, "Fine animal gorilla."

Dolphins

In 1979, Dr. Herman at the Kewalo Basin Mammal Laboratory in Hawaii began working to see if dolphins could learn a language. Why a dolphin? There are many reasons. Like humans, dolphins are mammals that feed and raise their young. They live in groups and rely on communication in catching food and defending themselves.

Perhaps the main reason is, like us, dolphins communicate mainly through sound. They produce high-pitched clicks, whistles, quacks, barks, and growls and are very good at learning new sounds. This is a big advantage in teaching them a language. But the real question is are dolphins smart enough to learn the meaning of words and the rules for putting them together? Many scientists think they are.

The dolphin language of sound is a lot different than the sounds we produce with our vocal cords. Producing "dolphin" sounds requires the use of creating different sounds, each with its own meaning, using a computer.

In order to test the ability of dolphins to learn a language, Dr. Herman used two mammals—Phoenix and Akeakami (Aki). Phoenix was taught using computer word sounds while Aki learned arm signal words. Even though they learned in different ways, they both were taught the same words. They learned nouns like *gate*, *window*, *water*, *person*, *hoop*, *frisbee*, and *fish*. They also learned words like *tail-touch*, *under*, *over*, *through*, *spit*, *fetch*, and *toss*. Of course the dolphins had to learn more than just the words; they also had to learn the rules for combining the words together.

After a great deal of training, each dolphin showed she could respond correctly to the language she learned. Phoenix and Aki could respond to sentences like "surface hoop tail-touch" and "bottom pipe fetch surface frisbee." Their actions showed that they understood the meaning of the words they were taught and could respond to words when used in new situations.

The next step in talking to dolphins may come with a computer-generated language. Maybe a computer can learn dolphin language and respond back with clicks, whistles, and growls.

So who knows? We'll probably never know what goes on inside an animal's head, but maybe in the future we will be able to go to the beach or zoo and carry on a conversation with the animals. I wonder what we'll say?

GA1512

I Didn't Know That!

Use the alphabet code below to decode the correct "animal answer" for the following statements.

A	B	C	D	E	F	G	H	I	J	K	L	M
1	2	3	4	5	6	7	8	9	10	11	12	13
N	O	P	Q	R	S	T	U	V	W	X	Y	Z
14	15	16	17	18	19	20	21	22	23	24	25	26

1. The American 16-18-15-14-7-8-15-18-14 can run at speeds of up to 60 miles per hour for two miles.

2. Young 11-1-14-7-1-18-15-15-19 are referred to as "joeys."

3. The smallest mammal is a 19-8-18-5-23 and weighs about as much as a dime.

4. The largest mammal is a 2-12-21-5 23-8-1-12-5 and can weigh as much as 170 tons.

5. The hump on the back of a camel is really made of 6-1-20.

6. 16-15-18-3-21-16-9-14-5-19 don't really "throw" their quills.

7. Many farmers do not care for 18-1-2-2-9-20-19 because sixty of them will eat as much plant life as a cow.

8. Some 2-1-20-19 hibernate more than half their lives.

9. New forests are "planted" when 19-17-21-9-18-18-5-12-19 plant acorns and other seeds.

10. The 1-18-13-1-4-9-12-12-15 has more than one hundred teeth—more than any other animal (except for certain whales).

11. The 13-15-12-5 has twenty-two bright pink "fingers" on its snout arranged like flower petals.

12. One-half of the total species of living mammals are classified as 18-15-4-5-14-20-19.

13. The "Black Death" during the Middle Ages was caused by 18-1-20-19.

14. The oil from a 19-11-21-14-11 is sometimes used in perfumes.

15. The 12-12-1-13-1 may carry a load of over one hundred pounds at altitudes which no other animal can tolerate.

Talking Back

Unlike most animals, much of the communication we do is in the form of words, and in our vocabulary there are lots of strange words. For example, some words spell the same word both forward and backward. These words are called palindromes. See if you can locate all the palindromes in the following story.

One morning when I woke up I knew that it was going to be a strange day. It was just one of those days. Anna came into my room and told me that baby Sis had taken the birthday present that Mom, Dad, and I had bought for Anna's birthday and then disappeared. Something about that story didn't seem on the level. I know I had hidden the present so well that I might not even be able to find it and besides, five minutes ago Sis was sitting at the table with food all over her bib.

A little after noon, my best friend Bob came over to see if I wanted a pup that his neighbor was trying to give away. I thought for a minute. I didn't have a present anymore, and it would be a good deed to take the dog. But I quickly decided that my parents might not appreciate a dog as much as Anna or I would. Besides, I had more important things to worry about like what happened to my little tot of a sister and what she did with the gift.

Bob said he would help me search for Sis if we could stop at the store and buy a can of pop. I said, "Sure, and maybe Bub will be there and he could help us look."

We ran to the store as fast as we could and Bob said, "Madam, we would like two cans of pop." We quickly drank them and were off again on our search.

We looked everywhere and would have been run over had we not heard the horn toot coming from a big car that zoomed past us. "I can't take this anymore. I'm too tired," Bob said. "Let's just send out an SOS and let the Coast Guard look for her."

I was getting tired, too, and it was getting dark, so I suggested that we just go home and guess what. I went into my little sister's room and there she was asleep, as quiet as a mouse, not even a peep. She had been there all the time. Anna had just wanted to get me out of the house so she could look for her birthday present.

"So, that's your game," I said. By now, Mom and Dad were home and had heard the whole story. "You've been on quite an adventure," Dad said. "I think you deserve a gift. And out from under his arms he pulled the little pup that Bob's neighbor had given them.

Even though the day did not start out very well, it had a happy ending, because that evening I had a new friend for life. I had a new dog I named Eve.

Learning Extenders

1. Create a musical piece (either instrumental or vocal) about communicating with animals.

2. Pretend that you are a marine biologist and are working on a project to communicate with dolphins. Keep a diary about your adventure and how you are going about "making contact."

3. Research how dogs learn and write a short "How-to" book titled, "You, Too, Can Teach an Old Dog New Tricks!"

4. Debate the subject, "Should humans learn to communicate with animals?" Make sure that you find good reasons why you are for or against this question.

5. Akeakami and Koko are two famous animals who were "taught to communicate." Research these animals and, in a short paper, compare how each was taught.

6. Locate several articles or books on animal communication from your school or community library. Construct a bibliography that other students might use. Provide a summary and review of at least three of the books or articles.

7. Jacques Cousteau is one of the most famous underwater explorers. Pretend you have been hired by the local newspaper to interview him about his efforts to communicate with animals. What would your story say?

8. Do dogs and cats talk to each other? Write a children's book on the topic of conversations dogs and cats might have.

9. Many people think that birds sing for our enjoyment. That could not be farther from the truth. They sing to attract a mate, to defend a territory, and to warn of danger. Tape record several bird songs and see if you and your classmates can tell what message they are trying to get across. You might need some help from a local college or university to identify the callings.

10. You have just been asked to write a twenty-word mini glossary of words important to the subject of animal communication. What words are you going to select and why? What do they mean? Consider words like *appeasement*, *gesture*, *colony*, *imprinting*, *instinct*, *mimicry*, *pheromone*, *stridulation*, and *symbiosis*.

11. Construct a class learning center on ways animals communicate. Be sure to include colors and patterns, smell, touch, sounds.

Developing Extenders for Academically Advanced Students

One of the major problems in classrooms is being able to meet the wide range of educational needs of students. While the chronological age in a classroom may be within one or two years, the abilities of students may range from five to six years or more. How then can teachers meet all the educational needs of the students?

It has been well documented that learning occurs in stages. Each of these stages has been defined and identified by Benjamin Bloom. Using the taxonomy Bloom developed, teachers can develop appropriate activities which will stimulate the thinking skills in all students at all levels. The six stages of cognitive development are:

Knowledge–remembering information
Comprehension–understanding information
Application–using information in new ways
Analysis–breaking down information into
its parts
Synthesis–putting information together in a
new way
Evaluation–judging the value of information

Many of the activities and learning extenders have been developed using "trigger" verbs at the various levels. However, please do not limit yourself or your students to just those given as a beginning. Work to develop your own personal activities which might better meet the needs and interests of the students in your classroom. The next three pages should assist you in that process.

The next two pages provide you with a simplistic version of Bloom's Taxonomy of Educational Objectives. You are given the definition, what the student does at that level, verbs associated with cognitive development at each level, and several sample product activities for each level. You have also been provided with a student Learning Extender Contract. This contract may be used in working with individual students in developing their own personal extender activities at the various cognitive levels. You and they can decide which level(s) is appropriate and work to develop an activity or activities that meet both the educational and interest needs of the student.

When working with academically talented students, be sure to encourage them to reach for that next level and not to be satisfied with just those things they do well. Also encourage them to use processing skills whenever possible. Ask them to observe, classify, infer, communicate, measure, predict, and experiment.

There are also some basic guidelines that should be followed when relating instructions to the characteristics of the gifted and talented student. These include:

1. Communicate the learning outcomes directly to the student.
2. Include only a minimum of repetition, review, and simple mnemonic devices.
3. Call for complex verbal and numerical responses and for original products, not merely recognition and recall.
4. Allow for simple responses to be made covertly.
5. Concentrate on student responses where there is some real challenge, some real need to think, and some real possibility for error.
6. Make sure that the "steps" or problems are large enough to prevent boredom.
7. Include strategies for varying the pace of instruction and learning.
8. Give opportunities for the student to judge his or her own responses and products.
9. Avoid excessive preoccupation with sequence and order. Give the student opportunities to organize.
10. If feedback is provided, be certain to allow a wide range of correct responses. Don't be preoccupied with "one right answer."
11. Provide opportunities for expression and development of attitudes and values as well as for knowledge and concepts.
12. Provide instruction which allows the student to progress steadily from simple problems, with a good deal of guidance, to much more complex problems, with a strong emphasis on self-directed inquiry.
13. Capitalize on the academically talented student's sense of humor. Be certain the materials are interesting, attractive, and enjoyable for the student.

Through the use of these suggestions, students will leave your classroom as individuals who have "learned how to learn." Through this effort, they can truly be their own teachers as long as they remain part of the environment.

Bloom's Taxonomy of Educational Objectives

Area of Taxonomy	Definition	What the Student Does	Verbs		Product Activities
Knowledge The instructor directs, tells, shows, examines, etc.	Recall specific information.	Masters the subject matter. Recalls bits of information. Knows dates, events, and places. Knows major ideas.	choose count define find group identify know label list	match memorize name read recite recognize record select write	Make a time line. List information. State main and related ideas. Make an alphabet book. Make a "fact" game.
Comprehension The instructor demonstrates, listens, questions, compares, contrasts, examines, etc.	Understands communicated material without relating it to other material.	Knows what the material says. Explains in own words. Illustrates ideas. Changes the form of ideas without changing the meaning.	associate change compare define describe estimate	explain extrapolate group reorganize simplify translate	Restate a paragraph. Draw a picture to show the main idea. Write/perform a play from the main ideas. Put similar ideas into categories. Make a rebus story. Make a coloring book for others.
Application The instructor shows, facilitates, observes, criticizes, etc.	Using methods, concepts, principles, and theories in new situations.	Applies what has been learned to another situation. Solves new problems. Illustrates and uses knowledge.	apply calculate choose classify construct employ examine experiment	illustrate interview model modify record select solve utilize	Apply learned information to new situation. Write about events related to the subject. Graph information. Make a mobile. Write a diary. Draw a map. Make a scrapbook. Make a model. Make a collage. Write a textbook for others.

Bloom's Taxonomy of Educational Objectives

Area of Taxonomy	Definition	What the Student Does	Verbs		Product Activities
Analysis The instructor proves, guides, observes, acts as resource, etc.	Breaking down a communication into its constituent parts.	Understands the various parts of what is being studied. Sees the parts of a subject as different entities. Develops a list of related facts or ideas about the subject.	analyze appraise calculate classify combine contrast deduct detect divide	examine group infer order relate separate simplify summarize test	Develop a related dictionary for a subject or area. Make a jigsaw/crossword puzzle. Construct a graph, chart, or diagram. Write a questionnaire. Complete a survey.
Synthesis The instructor selects, extends, analyzes, evaluates, etc.	Putting together constituent elements or parts to make a whole.	Does something new and different with learned information. Communicates learnings/experiences in creative way. Compares/contrasts information. Develops broad ideas from specific ideas.	arrange assemble build combine compose create design develop formulate	generalize imagine integrate invent organize plan predict produce specify	Write a TV/radio show. Make an invention/machine. Write a play/poem/song. Design something new. Combine two objects to create something new. Write a news article. Sell or advertise an idea.
Evaluation The instructor accepts, harmonizes, etc.	Judging the value of materials and methods.	Judges the value of ideas, purposes, methods, etc. Judges the accuracy of ideas. Supports own ideas against the arguments of others.	appraise assess choose critique decide determine estimate evaluate grade judge	measure rank rate recommend revise score select solve test	Conduct a court trial. Keep an "impression" diary on subject studied. Hold a debate. Write a business letter. Evaluate items and tell why you did/did not like them. Hold panel discussion. Draw conclusions.

Learning Extender Contract

Topic	Knowledge	Comprehension	Application	Analysis	Synthesis	Evaluation
	choose memorize count name define read find recite group recognize identify record label select list write match	associate explain change extrapolate compare group define reorganize describe simplify estimate translate	apply illustrate calculate interview choose model classify modify construct record employ select examine solve experiment utilize	analyze examine appraise group calculate infer classify order combine relate contrast separate deduct simplify detect summarize divide test	arrange generalize assemble imagine build integrate combine invent compose organize create plan design predict develop produce formulate specify	appraise judge assess measure choose rank critique rate decide recommend determine revise estimate score evaluate select grade solve

Bibliography/Resources Used

____

____

____

____

Criteria for Evaluation

	Student Evaluation	Teacher Evaluation
Worked well independently	__________	__________
Used wide variety of resources	__________	__________
Used appropriate resources	__________	__________
Managed time and resources well	__________	__________
Arranged for completion/sharing of project	__________	__________
Completed contract as agreed upon	__________	__________
Other________________	__________	__________

GA1512

Answer Key

Rock and Roll Page 3
MEGALITH

And the Plot Thickens! Page 4
THE STONE AGE

You Hide Your Age Well! Page 7

```
A G W M S C P B A G E L N
G A S E G A Y O V A W G A
T U G S P I L L A G E M V
V G E G A W O T P E G A B
I L E M R G S C O R S A G E
L R A G E M R A L E W E G
L E G A T S E U E W G G A
A G E N G A G E W A S E T
G A G U E G A E L G M A L
E L E G A E L I M E G A O
S L A G G E S R V S W M V
T I E G A C A B I O P G S
V T B A G L G S R C S V A
```

Who's First? Page 8
movie projector, holography, color television
bicycles, elevator, airplane
sewing machine, cash register, adding machine,
  pocket calculator
stethoscope, hypodermic syringe, aspirin, X ray

ESP and Other Abbrev. Page 12
HOW PERCEPTIVE U R!

Memory Overload Page 28
1. 5
2. second
3. 3:00
4. polka dots
5. 4751
6. 2
7. 3
8. yes
9. no
10. ice cream
11. two
12. 2
13. yes
14. City St.
15. Craig's Cola
16. no
17. 27

Memory Makers Page 29
WHAT A MEMORY!

In Your Dreams Page 33
[(434 + 630) - 324] - 45 - 525 = 170
(2570 - 140) - (42 x 3.75) = 2272.5

Author, Author! Page 34
UNOPENED

Take Flight! Page 38
AN ASTRONUT

It's as "Plane" as the Nose on Your Face Page 39
1. lone, loan
2. steal, steel
3. earn, urn
4. peak, peek
5. sent, scent
6. aloud, allowed
7. rowed, road
8. fair, fare
9. pane, pain
10. blew, blue
11. whole, hole
12. bored, board
13. heard, herd
14. real, reel
15. seize, seas

A Compound Problem! Page 43
AN EVERCHANGING VIEW

Don't Bug Me with All Your Questions! Page 48
An Aunt Hill
Katydid
The Newlywebs

I Didn't Know That! Page 52
1. Pronghorn
2. Kangaroos
3. Shrew
4. Blue whale
5. Fat
6. Porcupines
7. Rabbits
8. Bats
9. Squirrels
10. Armadillo
11. Mole
12. Rodents
13. Rats
14. Skunk
15. Llama

One morning when I woke up ______ y. It was just one of those days. **Anna** came into my ro______ e birthday present that **Mom** and **Dad** and I had bou______ d. Something about that story didn't seem on the **level**______ I might not even be able to find it and besides, five mi______ all over her **bib**.

A little after **noon**, my best fr______ at his neighbor was try- ing to give away. I thought ______ and it would be a good **deed** to take the dog. But I c______ eciate a dog as much as **Anna** or I would. Besides, I ______ e what happened to my little **tot** of a sister and what ______

Bob said he would help me s______ buy a can of **pop**. I said, "Sure, and maybe **Bub** will b______

We ran to the store as fast a______ ike two cans of **pop**. We quickly drank them and wer______

We looked everywhere and w______ horn **toot** coming from a big car that zoomed past ______ **ob** said. "Let's just send an **SOS** and let the Coast Gu______

I was getting tired, too, and______ just go home and guess what. I went into my little s______ as a mouse, not even a **peep**. She had been there a______ out of the house so she could look for her birthday p______

"So, that's your game," I sa______ d heard the whole story. "You've been on quite an ad______ And out from under his arms he pulled the little **pup**______

Even though the day **did** no______ cause that evening I had a new friend for life. I had a ______